THE MOLECULAR BIOLOGY OF CELL MEMBRANES

THE MOLECULAR BIOLOGY OF CELL MEMBRANES

PETER J. QUINN
Department of Biochemistry,
Chelsea College, University of London

UNIVERSITY PARK PRESS
Baltimore London Tokyo

First published 1976 by
The Macmillan Press Ltd
London and Basingstoke
Associated companies in New York Dublin
Melbourne Johannesburg and Madras

Published in North America by
University Park Press
Chamber of Commerce Bldg.
Baltimore, Maryland 21202

Library of Congress Cataloging in Publication Data

Quinn, Peter J.
The molecular biology of cell membranes.

Includes bibliographical references and index.
1. Cell membranes. 2. Molecular biology.
I. Title. [DNLM: 1. Cell membrane. 2. Molecular biology. QH601 Q7m]
QH601.Q56 1976 574.8′75 76–5817
ISBN 0–8396–0929–6

Printed in Great Britain

Contents

Preface

The past few years have witnessed astounding advances in our appreciation of the structure of cell membranes and the molecular mechanisms underlying their different functions and properties. The purpose of this book is to provide students and teachers involved in various disciplines of biology with a contemporary guide to these developments. Special emphasis has been placed on the type of molecules found in cell membranes, on how they are arranged in the structure, and on the way membranes assist in regulating different cellular processes. The content and presentation should be within the grasp of undergraduates with an elementary knowledge of biology, chemistry and physics but an adequate grounding in the first should be sufficient to comprehend most of the text.

The first chapter comprises an outline of the distinguishing features of the different membranes found in animal cells together with a description of their particular chemical and biochemical characteristics. The principal objective in dealing with membrane structure has been to relate the chemical composition of individual membrane constituents to their intrinsic physical properties and to examine the nature of the interactions between these components in the assembled membrane. For this reason, a chapter (chapter 2) has been devoted to the physical structure of membrane proteins and lipids, including an extended discussion of the associative behaviour of lipids dispersed in aqueous systems, with a further chapter (chapter 3) attempting to draw this information into a coherent picture of how molecules are organised in biological membranes. Molecular principles have been retained as far as possible when describing the mechanisms of membrane permeability and transport (chapter 4) and other specialised properties and functions of particular cell membranes (chapter 5).

It is perhaps inevitable that the major advances in our knowledge of the molecular architecture of cell membranes has been achieved, in the main, by the development and adaptation of a variety of sophisticated physical techniques especially for this purpose. I believe it is appropriate to include some coverage of this aspect of the subject, although, in deference to students lacking an advanced training in physics and chemistry, the treatment has been restricted to a brief outline of the principles involved and the type of information that can be derived from a particular technique. References are provided to aid those who need further information about this aspect.

The text originated, in part, from a course given to undergraduates in biochemistry at Oxford and I am grateful to my students for their inquisitiveness and enlightening criticisms. My own approach to the study of cell and model membranes has been greatly influenced by working in a number of stimulating research groups both in this country and abroad; I am therefore indebted to my colleagues for the inspiration they have given me, but at the same time I must claim responsibility for any ambiguous or even erroneous material that may still remain. I would like to give special mention to Charles Pasternak and members of the Oxford research group for the invaluable assistance they have given me.

Finally, and by no means least important, there has been the support I have received from my wife and children, who cheerfully sustained my efforts throughout untold difficulties.

Oxford P. J. Q.

1 Characterisation of Animal Cell Membranes

1.1 Cytological Features

Because of their small size most subcellular organelles cannot be resolved by visible-light microscopy, although finite boundaries of cells in suspension can be deduced from changes in shape and size when the ionic strength of the suspending medium is altered. When cells are damaged mechanically by, for example, a precipitous exposure to media of low osmotic strength or by freezing, the cell contents leak out and the cells ultimately lose viability. These and other observations have clearly established that large molecules such as proteins, DNA and certain dyes of high relative molecular mass are incapable of penetrating the intact cell membrane. Further studies have examined the permeability characteristics of a range of much smaller inorganic ions and organic non-electrolytes, and it has been discovered that the membrane is remarkably selective with respect to those molecules it allows to penetrate and those it does not. The basis of this selectivity was found to be markedly dependent on the chemical structure of the permeant molecules as well as on their size: molecules possessing lipid-like or lipophilic properties tend to penetrate more readily than hydrophilic solutes and ions. This feature implies that the membrane possesses physical properties consistent with a region distinctly hydrophobic in character.

Many important revelations of cellular ultrastructure accompanied the introduction of the electron microscope and steady progress has been made since then in improving techniques for the preparation and staining of thin tissue sections for examination. The existence of a number of subcellular organelles such as the Golgi complex and the endoplasmic reticulum were in fact unknown before the advent of the electron microscope. Membranes in stained thin section, as visualised by the electron microscope, appear as trilamellar structures with two outer electron-dense layers sandwiching an intermediate lucent layer. Differences in the fine structure of subcellular membranes are sometimes discernible but the usual method of identifying particular membranes is to use the location of the membrane within the cell and the morphology of the organelle with which it is associated. Illustrative guides to the fine structure of various animal cells and tissues can be found in texts by Fawcett[1] and Porter and Bonneville[2].

1.1.1 The Plasma Membrane or Plasmalemma

The plasma membrane constitutes the cell boundary and in electron micrographs the membrane exhibits a typical trilamellar configuration. Each band is approximately 3 nm in width and the overall dimension is in the order of 8 nm to 12 nm. The outer surface is frequently coated with a diffuse layer, which has been shown by histochemical techniques to be rich in polysaccharide. In order to allow certain cells to perform particular functions, their plasma membrane may differentiate to form specialised structures. These structures may be transient and continually changing or they may be stable and of recognisable form. The plasma membrane of *Amœba*, for example, undergoes transient differentiations in the form of large projections or pseudopodia on which the organism relies for its ability to move and ingest extra-cellular food particles. The latter process is referred to as phagocytosis and is common to specialised cells of higher organisms such as polymorphonuclear leucocytes and macrophages, cells which constitute important components of the body's defence system against invasion of foreign organisms. During phagocytosis the engulfed material is eventually pinched off from the plasma membrane to form a membrane-enclosed vacuole in the cytoplasm. A closely related process called pinocytosis or endocytosis is a feature of many different cell types but in this case relatively smaller volumes of usually non-specific extracellular material are incorporated by means of slender pseudopodial projections. Occasionally very small vacuoles (less than 200 nm in diameter) may be observed, which arise from minute invaginations of the plasma membrane surface. These may either remain attached to the cytoplasmic surface of the membrane or they may form spiny projections and become detached from the membrane.

Some epithelial cells, such as those of the intestinal mucosa and proximal renal tubules, possess stable differentiations of the plasma membrane. In these cells the membrane forms protrusions called microvilli, which are regular in

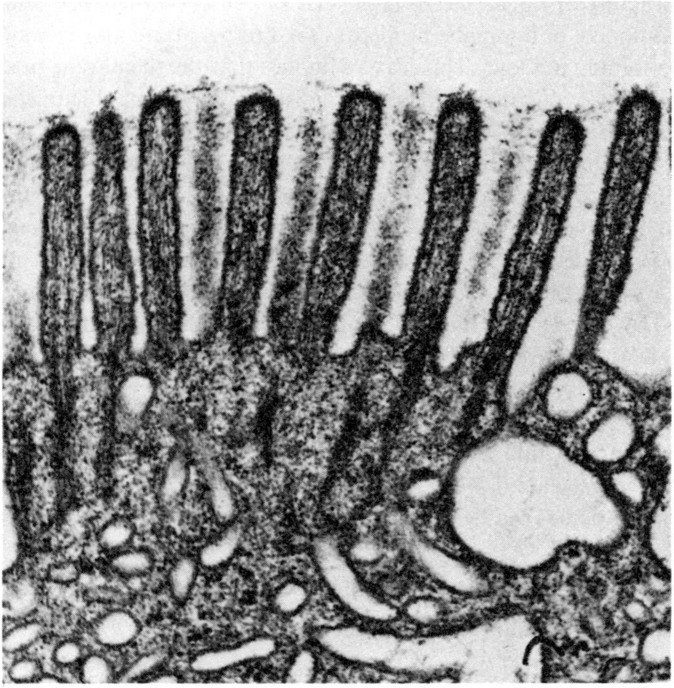

Figure 1.1 Electron micrograph of a thin section through intestinal microvilli of suckling rat ileum. Magnification × 35 000 (micrograph by S. Knutton).

shape and serve to increase greatly the surface area of these cells exposed to the lumen (see figure 1.1). Another stable structure formed from the plasma membrane is the myelin sheath surrounding certain nerve axons (see figure 1.2). The membrane forming the sheath arises from another specialised nerve cell, the Schwann

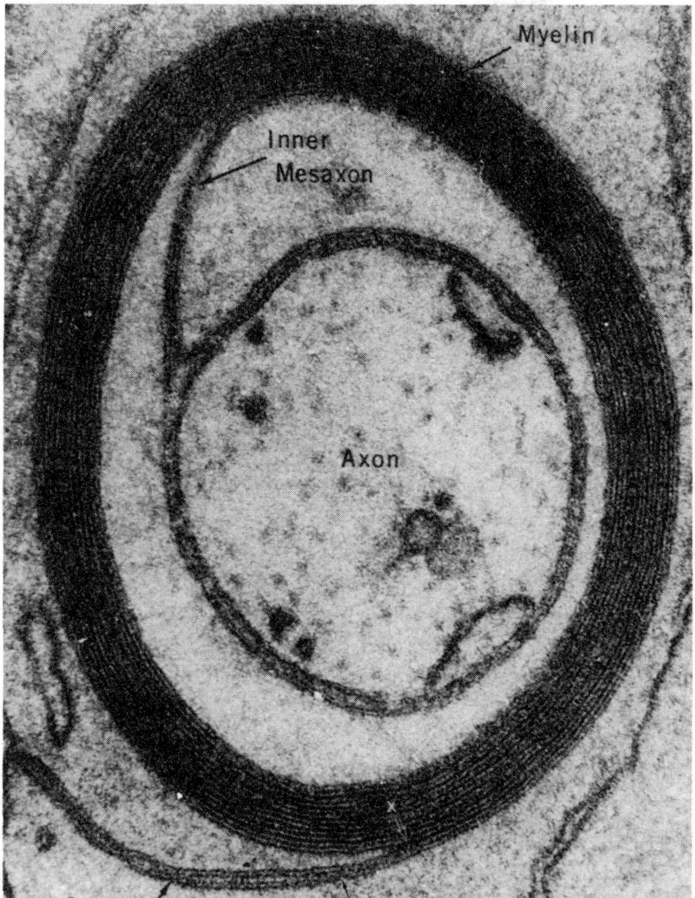

Figure 1.2
(a) An electron micrograph of a myelinating fibre of mouse sciatic nerve sectioned normally to the axon. The preparation was fixed with potassium permanganate. The arrow, x, indicates where the cytoplasmic surfaces of the Schwann cell plasma membrane align to form the denser line of the 12 nm myelin period, and at y the outer surfaces of the membrane forming the outer mesaxon are seen coming together to form the less dense interperiod line. Magnification × 13000 (micrograph by J.D. Robertson).

cell, which is closely aligned with the nerve fibre. During myelination, the Schwann cell plasma membrane becomes wrapped around the axon, forming a large number of concentric layers of membrane in close apposition. Figure 1.2a shows an electron micrograph of a peripheral nerve during myelination and the process is illustrated schematically in figure 1.2b.

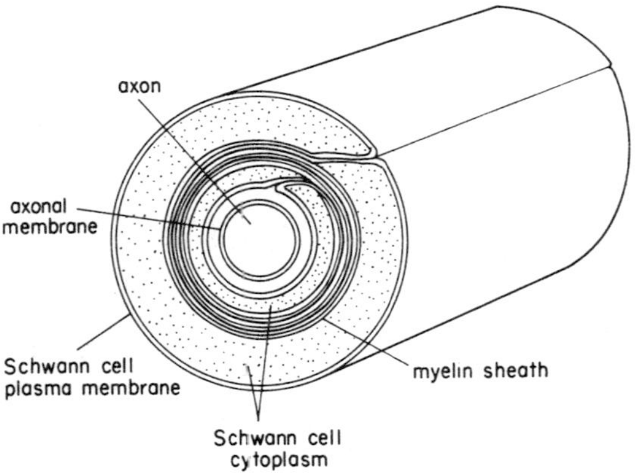

Figure 1.2
(b) A diagrammatic representation of peripheral nerve myelination showing the plasma membranes wrapping around the nerve axon.

1.1.2 Nuclear Envelope

All eukaryotic cells have a nucleus surrounded by a membrane envelope. Eukaryotic cells can be distinguished from prokaryotic cells because the latter do not have a specialised membrane enclosing the chromatin. The nuclear envelope (see figure 1.3a) consists of two membranes each approximately 7.5 nm thick and separated from each other by a space varying from 40 nm to 70 nm. The two membranes join or anastomose at intervals to form nuclear pores about 80 nm in diameter (see figure 1.3b). These pores do not permit free passage of ions or amino acids between the cytoplasm and the nucleoplasm. However, on the basis of electrical-potential measurements, their movement is likely to be less restricted than passage directly through the membrane[3]. Markovics *et al.*[4]

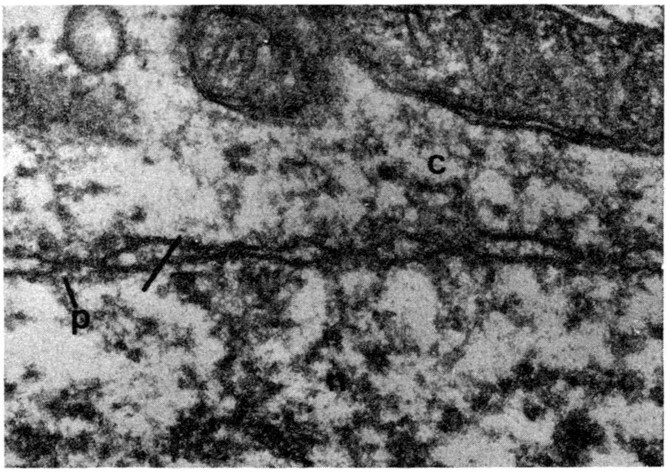

Figure 1.3a
(a) Electron micrograph of the nuclear envelope of rabbit Deiters' Neuron sectioned normal to the surface of the membranes. Note that the inner and outer membranes of the envelope are continuous with each other at the nuclear pores (p). The nucleoplasm (n) and cytoplasm (c) are indicated. Magnification × 77000 (micrograph by J. Metuzals).

examined the distribution of pores in nuclear membranes from a variety of cell types and found them to be more or less randomly distributed when the cells were in a growth phase but that they tended to cluster during the remainder of the cell cycle. During cell division the nuclear envelope disappears and is thought to be reassembled during telophase from elements of the endoplasmic reticulum.

1.1.3 Endoplasmic Reticulum

As its name implies, this is a system of membranes that form cavities or channels extending throughout the cytoplasm. The membranes may form extended tubules about 50 nm in diameter or can be arranged in sheetlike structures, which, in thin section, appear as paired trilamellar membranes enclosing a space varying from 30 nm to 60 nm. A number of similarly oriented structures

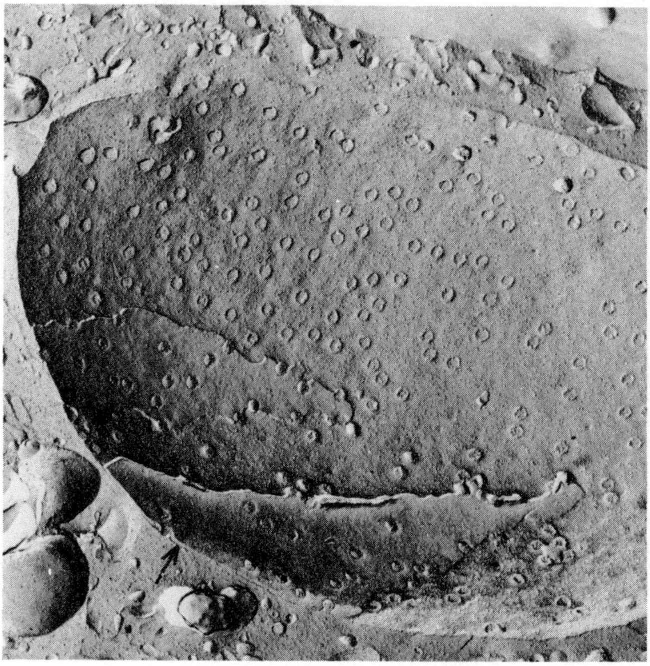

Figure 1.3b
(b) A freeze-fracture replica of the nuclear envelope of a P815Y mouse mastocytoma cell showing the distribution of nuclear pores. The density of membrane-associated pores is lower in the fracture face of the juxtanuclear membrane; the intermembrane space is indicated by the arrow. Magnification × 13600 (micrograph by S. Knutton).

frequently associate and so form a lamellar system. The total amount of endoplasmic reticulum observed in the cytoplasm and the organisation of the constituent membranes varies from one cell to another but generally the extent of its development reflects to some degree the metabolic activity of the cell.

Endoplasmic reticulum can be classified into two forms depending on the presence or absence of ribonucleic acid granules attached to the surface of the membrane[5] (see figure 1.4). The two forms are known as rough-surface and smooth-surface endoplasmic reticulum, respectively. Certain proteins are synthesised on membrane-attached ribosomes and when cellular-protein biosynthesis increases, the proportion of rough-surface endoplasmic reticulum relative to the smooth variety increases. The smooth-surface endoplasmic reticulum appears to be concerned with drug detoxification and the amount of this type of membrane can be increased up to tenfold in liver cells following the administration of certain barbiturate drugs. The newly synthesised membrane is indistinguishable from the original membrane with regard to morphological characteristics but its protein and enzymic constituents differ in their proportions. It is noteworthy that the total amount of rough-surface membrane under these conditions remains relatively constant.

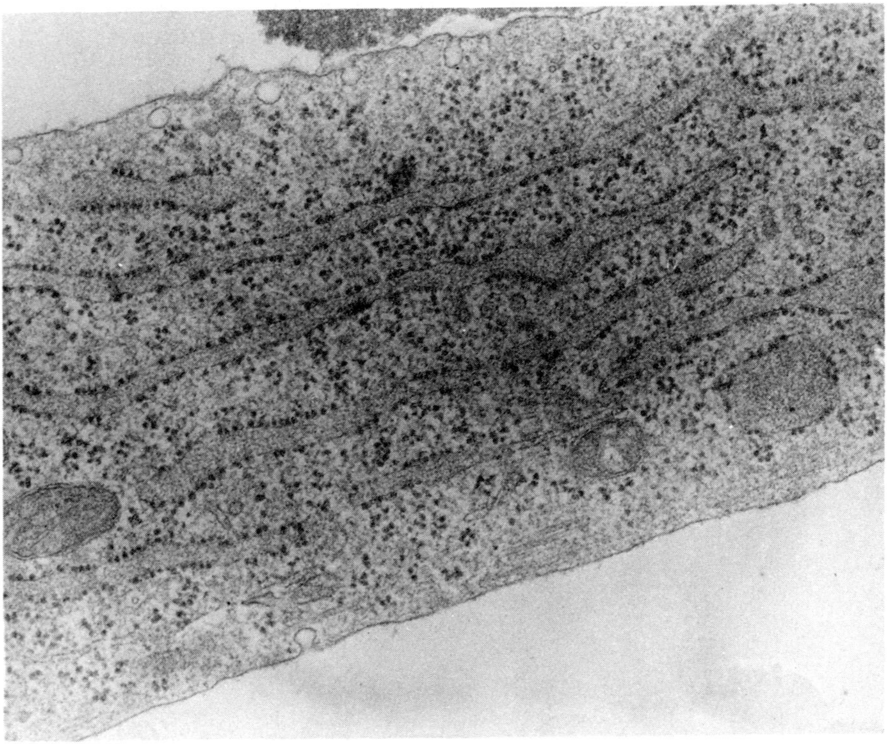

Figure 1.4 Electron micrograph of rough-surface endoplasmic reticulum of an SV40 transformed 3T3 mouse fibroblast cell showing ribosomes attached to the membrane. Magnification × 10300 (micrograph by S. Knutton).

The outer membrane of the nuclear envelope is occasionally continuous with the endoplasmic reticulum and ribosomes may also be attached to it. The smooth-surface endoplasmic reticulum is generally arranged in a tubular form rather than bounding cysternae and it is often found associated with the Golgi complex from which it is difficult to distinguish.

1.1.4 The Golgi

The Golgi apparatus or complex is usually located in a region of the cell near the nucleus and, as seen in figure 1.5, consists of a stack of usually four to eight

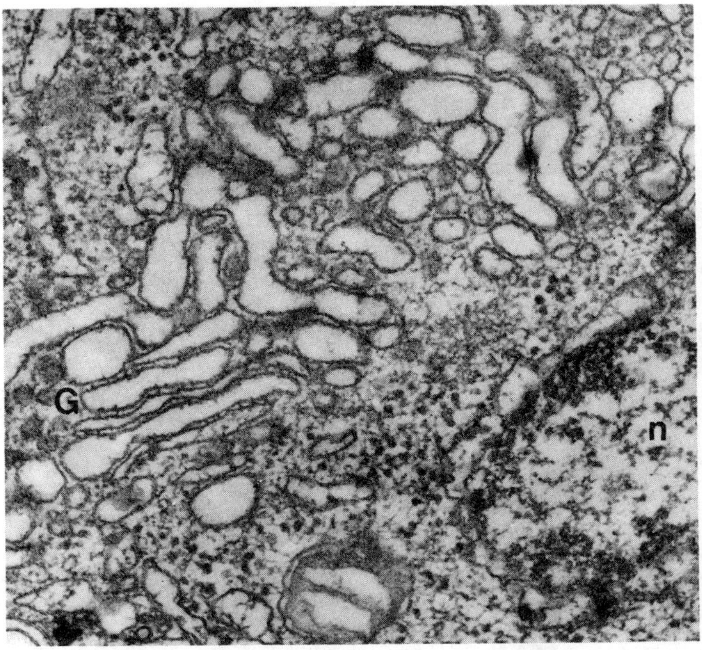

Figure 1.5 Electron micrograph of a thin section through a rat intestinal epithelial cell showing the Golgi complex (G). The nucleus (n) and the surrounding envelope can be seen on the left. Magnification × 21 500 (micrograph by S. Knutton).

membrane-enclosed cysternae in close parallel array. The cysternal space extends over a distance of about 15 nm. Small vesicles of between 40 nm and 80 nm in diameter, which are apparently pinched off from the periphery, can be observed at the ends. The innermost cysternal space of the complex is often swollen and closely associated with vacuoles found to contain products intended for secretion from the cell[6].

1.1.5 Mitochondrial Membranes

Mitochondria located in the cytoplasm can vary in shape[7], but are usually cylindrical structures about 3 μm in length although in some cells they may be as long as 10 μm. A specialised form of mitochondrion is found in spermatozoa: here they are coiled, end to end, along the entire length of the principal piece of the tail. Mitochondria possess two distinct membranes (figure 1.6): an outer smooth membrane about 7 nm thick and an inner membrane separated from the limiting membrane by a space of approximately 8 nm. The inner membrane has infoldings of variable length known as cristae and these projections form incomplete transverse or occasionally longitudinal septa through the matrix. In some cases the cristae are arranged in a tubular form and may connect the inner membrane on both sides of the mitochondrion. A certain amount of fine structure of the cristal membrane is revealed by negative staining of unfixed mitochondria. This technique involves treatment of mitochondria with stains, such as phosphotungstic acid, which permeate the aqueous regions of the organelle and thus provide a contrast with membranes, which do not take up the

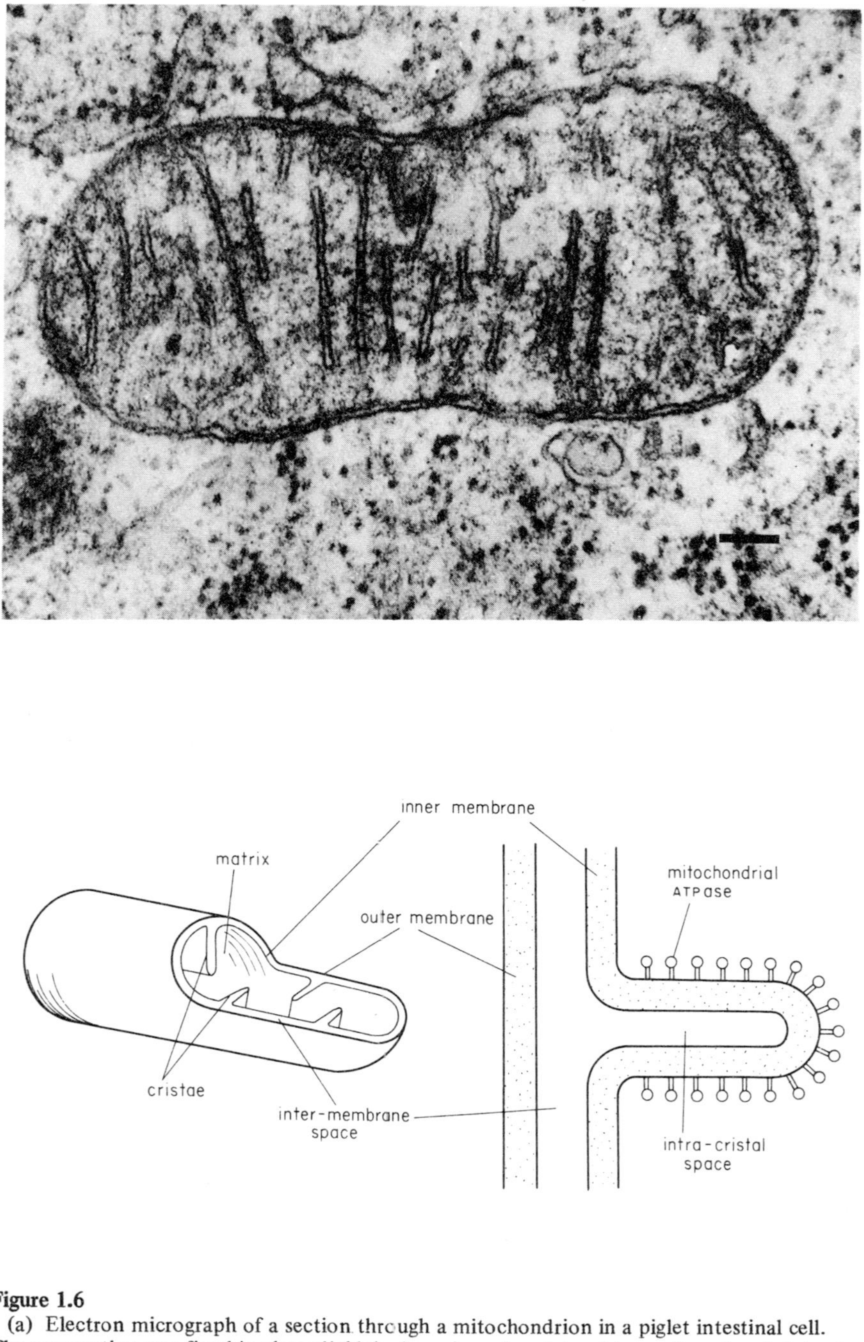

Figure 1.6

(a) Electron micrograph of a section through a mitochondrion in a piglet intestinal cell. The preparation was fixed in glutardialdehyde and osmium tetroxide, sectioned and stained with uranyl acetate and lead acetate. Magnification × 78000 (micrograph by E.A. Munn).

(b) Diagrammatic representation of the arrangement of mitochondrial membranes and the definition of mitochondrial compartments.

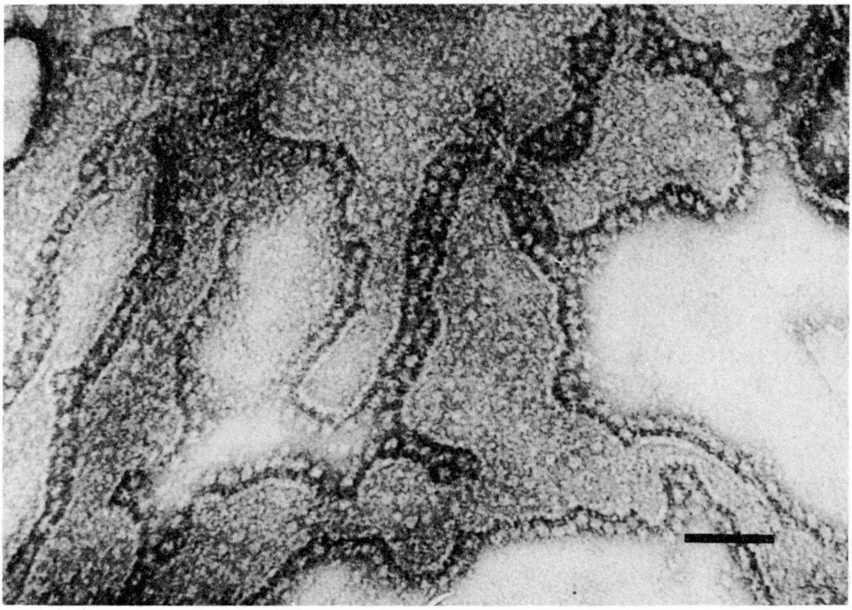

Figure 1.7 Electron micrograph of fragments of inner membrane from rat liver mitochondria. The preparation was negatively stained with 2 per cent sodium phosphotungstate. Bar = 50 nm (micrograph by E.A. Munn).

stain. An electron micrograph of such a preparation (figure 1.7) reveals small particles of about 9 nm diameter projecting from the matrix surface and these are connected to the membrane by a stalk, 4 to 5 nm in length. The density of particles varies between 2000 and 4000 per μm^2. There is some uncertainty about the changes in appearance of the membrane that accompany the negative-staining process, especially since positive staining of the membrane in thin sections of mitochondria with osmium, lead or uranyl compounds does not suggest the presence of such particles.

1.1.6 Other Membrane Inclusions

Membrane-enclosed particles are often observed in the cytoplasm and these include lipid droplets, pigment granules and lysosomes. The lysosomes are morphologically heterogeneous and can be distinguished from other membrane inclusions because they preferentially stain with heavy-metal ions such as tungstate; they are known to contain certain hydrolytic enzymes such as acid phosphatase, ribonuclease and cathepsins. Lysosomes of polymorphonuclear leucocytes are sometimes seen to fuse with ingested vacuoles and presumably enzymes concerned with the digestion of the vacuolar contents are liberated.

1.1.7 Structural Considerations in Thin Tissue Sections

All the morphological features described so far have been obtained by the now classical techniques of fixing, sectioning and staining tissue preparations, and such methods have obviously made a significant contribution to our understanding of the organisation of subcellular membranes in the various organelles. Electron microscopy is also a valuable adjunct to the isolation of subcellular membranes, since it provides an objective criterion for estimating the amount and often the source of contaminating material. A number of biochemical processes including the manner in which proteins destined for export are synthesised by secretory cells and the method whereby they are processed during passage through the cell have been worked out largely by refinements of the method (see section 5.2.4). Nevertheless, there are some serious limitations concerning the interpretation of electron micrographs, particularly with regard to the definition of molecular organisation of the constituents within membranes. This does not result from any deficiency in the resolving power of the electron microscope, which for practical purposes can distinguish detail down to less than 0.5 nm, well within the range required for most biological macromolecules, but is due to a combination of two basic uncertainties[8]. The first is the possibility that not all essential biochemical features are faithfully preserved during the preparation of tissues for examination and the second relates to difficulties in determining the precise location of heavy-metal atoms used to improve electron-density contrast in the sample.

It is well known that chemical fixation of tissues causes a substantial alteration in the conformation of membrane proteins and the most efficient fixing agents probably give the best definition to the trilamellar arrangement of membranes. Moreover, some preparative techniques often lead to the appearance of structural detail not observed when tissues are prepared by alternative methods; an excellent example of this effect is the presence of particles on the matrix surface of mitochondrial cristae membranes in negatively stained but not conventionally fixed membranes. The second factor, however, is considered to be the more serious when attempting to interpret molecular structure of membranes from ultrastructural appearance. Even though some differences in symmetry or homogeneity of the characteristic trilamellar arrangement can often be detected, there are as yet no objective criteria for establishing the precise nature of these features. Perhaps the most convincing evidence is derived from the persistence of the trilamellar appearance of myelin and mitochondrial membranes when they are stained after lipids have been extracted with organic solvents. Also, films formed from protein alone as well as aqueous dispersions of purified phospholipids all appear as trilamellar structures in stained section.

1.1.8 New Approaches in Electron Microscopy

Many of the difficulties encountered in chemical fixation of tissues for electron microscopy can be avoided by resorting to alternative methods of fixing tissues. A physical-fixation method relying on rapid freezing procedures developed many years ago[9] is a common method. Briefly, this entails rapidly cooling the specimen to below $-100°C$ so that the constituent molecules are 'immobilised' in their original state. The cooling rate must be fast enought to prevent the formation of ice crystals, which cause physical disruption of molecular arrangement as well as disturbance of electrolyte and proton distribution. Ice-crystal formation is often reduced by employing antifreeze agents such as glycerol or dimethylsulphoxide which, by themselves, do not cause perturbations in the structure. Cooling rates of up to 10^6 K s^{-1} have been claimed although even this may not be fast enough to preserve all the original molecular configurations when the rate of movement of some membrane molecules is taken into account. Nevertheless, there are certain grounds for encouragement in this regard. X-ray measurements of intermolecular distances between membrane components after freezing, for example, show that these remain the same as molecules at normal temperatures. Moreover, micro-organisms retain almost complete viability after freezing, suggesting that there are no irreversible structural changes associated with the freezing process.

Once frozen and placed in a vacuum chamber, the specimen is fractured mechanically and molecular detail from within the sample is revealed along the cleavage surfaces (see figure 1.8). Some areas of the exposed fracture face may be covered with a layer of ice and this may be removed by sublimation to reveal the underlying structure. This process is known as etching. The fracture face cannot be viewed directly under the electron microscope so it is necessary to prepare a replica of the surface. This is done by applying a thin deposit of a heavy metal such as platinum to the exposed surface and supporting this with a second, thicker layer of electron-translucent carbon particles. The sample is dissolved away and the replica is then ready for examination under the microscope. The structural detail obtained from replicas prepared by this method is not as good as that obtained by conventional thin-sectioning techniques, but for routine preparations resolution in the order of 2 nm can be achieved.

If membranes are exposed on the fracture face, the most interesting aspect, from a morphological viewpoint, is to be found when cleavage has occurred in a longitudinal direction: in this case it will always have done so along a central plane within the membrane[10]. When complementary fracture faces of a membrane cleaved in this way are etched, no further detail can be developed, which suggests that this region is devoid of excess water and is hydrophobic in character. These areas appear mainly smooth and are formed by the hydrocarbon region of the membrane lipids. However, small particles about 10 nm are also usually observed, and these are believed to represent interpolated membrane proteins, an interpretation supported by the fact that they can often be removed from the fracture face by treatment of the membranes with protease enzymes such as pronase[11], whereas phospholipase treatment simply increases the particle density (see section 3.1.1). In the human erythrocyte membrane these particles range in size from 8 nm to 14 nm in diameter and protrude 2 nm to 4 nm above the plane of the fracture face. The number and distribution of particles vary characteristically in different membranes and an asymmetric distribution between the two fracture faces of the membrane is commonly observed. The outer leaflet of the human red-cell membrane, for example, has a particle density of about 2500 per μm^2, whereas on the inner leaflet (juxtacyto-plasmic) the particle density is only about 600 per μm^2. In bladder epithelial plasma membranes, particles are present only on the inner surface of the outer leaflet; the surface corresponding to the cytoplasmic leaflet is smooth in appearance and free from particles. For further information on the application

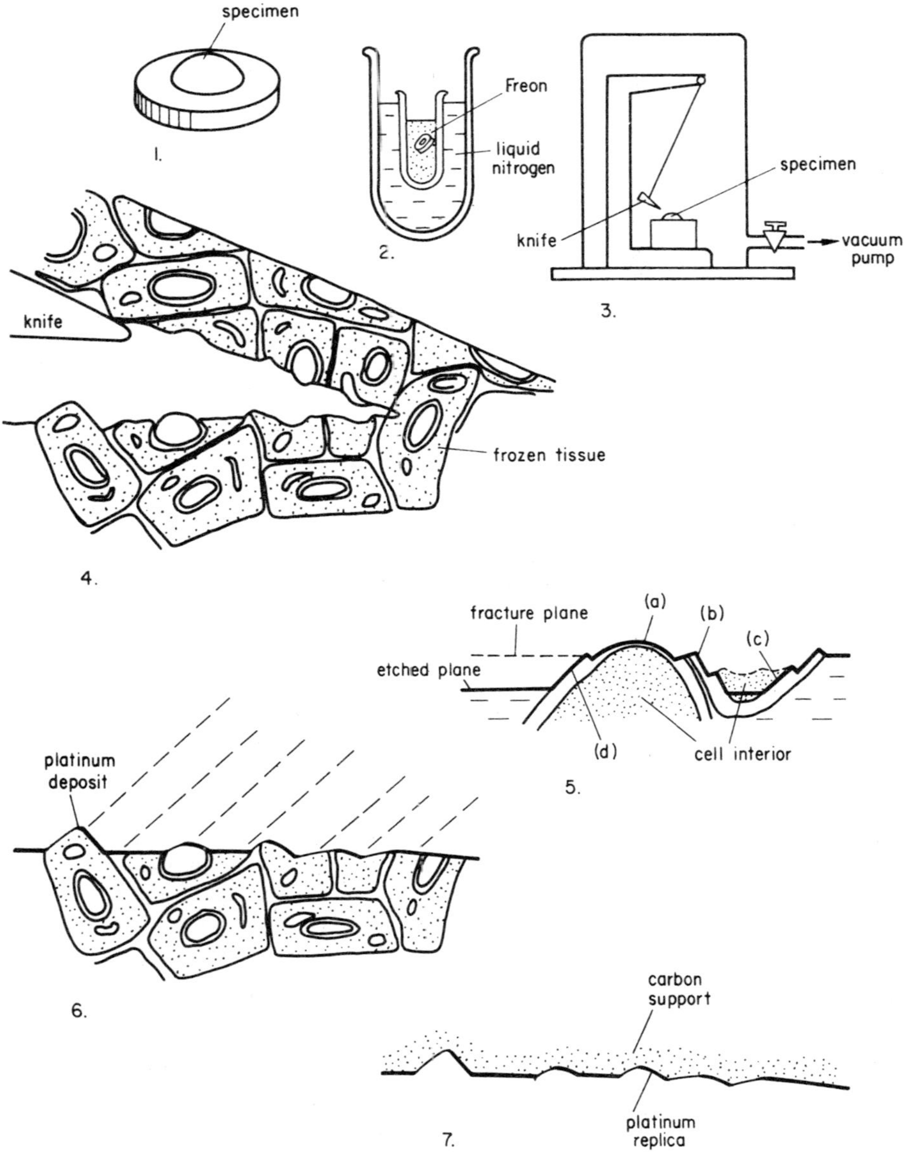

Figure 1.8 Freeze–cleave procedure. The sample (1) supported on a copper disc is rapidly frozen in liquid Freon to – 196°C (2). The frozen specimen is transferred to a vacuum chamber (3) and fractured with a cooled knife blade (4). The fracture face (5) exposes membranes cleaved along a central plane and reveals the inner surfaces of the inner (a) and outer leaflets of the membrane. When the water table is lowered by etching, the cytoplasmic (c) and external surfaces (d) of the membrane are revealed. The surface is shadowed with platinum at an angle to the fracture plane (6) and a second layer of carbon particles applied to support the metal film. After the sample has been dissolved away (7) the carbon–platinum replica is ready for viewing under the electron microscope.

of freeze–cleave techniques to study cell-membrane structure the reader should consult a recent review by Muhlethaler[12].

1.2 Biochemical Characteristics

Nearly all biological membranes possess some form of enzyme activity. Many of these membrane-bound enzymes are found only on one type of membrane and the localisation or assay of this activity provides a useful biochemical marker for a particular membrane[13,14]. Cytochemical techniques have been developed for a number of these marker enzymes and such methods have been used to study the location of membranes within the ultrastructure of the cell. An essential prerequisite for purifying membranes on the basis of enzyme-specific activity is that a particular enzyme must first be shown to be associated exclusively with one type of membrane. Some of the enzyme activities commonly used for purifying subcellular membranes of rat liver are presented in table 1.1. It must be emphasised that the example of rat liver represents a special case in which the tissue consists essentially of a single cell type. Most tissues are, in fact, a heterogeneous association of different cell types with varying resistance to destruction by homogenisation procedures. It is therefore mandatory in most cases to select additional criteria, such as morphological or cytochemical features, to establish the true identity of particular membrane fractions.

Table 1.1 Enzyme markers commonly used to characterise rat liver membranes

Membrane	Enzyme	Enzyme commission nomenclature	Number of different activities reported
plasmalemma	5′-nucleotidase	E.C.3.1.3.5	24
nuclear membrane	cytochrome c oxidase*	E.C.1.9.3.1	
endoplasmic reticulum	glucose 6-phosphatase	E.C.3.1.3.9	75
smooth-surface	uridine diphosphatase	E.C.3.6.1.6	
rough-surface	various esterases	E.C.3.1.1.1	
mitochondria			70
inner membrane	succinate dehydrogenase	E.C.1.3.99.1	21†
outer membrane	kynurenine hydroxylase	E.C.1.14.1.2	13†
Golgi complex	UDP-galactose: N-acetylglucosamine galactosyltransferase	E.C.2.4.1.22(A)	
lysosomal enzymes	acid phosphatase and other hydrolytic enzymes		14

*extra mitochondrial

†9 enzymes are common to inner and outer membranes

Biochemical differences between different membranes give rise to unique immunological specificities, which have been exploited to characterise various intracellular membranes. De Heer *et al.*[15], for example, demonstrated that each intracellular membrane of rat-liver cells, including nuclear membranes, endoplasmic reticulum, inner and outer mitochondrial membranes and lysosomal membranes possesses distinctive and unique autoantigenic determinants. No such determinants were detected on plasma membranes and furthermore none of the antigens appeared to be common to more than one membrane.

1.2.1 Membrane-bound Enzymes

Many enzymes associated with membranes are interpolated so that they form an integral part of the structure. These so-called membrane-bound enzymes can be distinguished from enzymes that may appear to adhere to the surface because they cannot be detached from the structure by repeated washing with physiological media. Procedures have been developed to extract membrane-bound enzymes but most of these involve the destruction of membrane integrity. These include the use of various ionic or nonionic detergents under neutral pH conditions, but often after adjusting the system to either high or low pH. An alternative method consists of removing lipid material from the membrane either with organic solvents or by digestion with phospholipase enzymes, followed by subsequent solubilisation of the residual protein with detergent when this is required. Such procedures, as one might expect, invariably sacrifice some enzyme activity, since organic solvents tend to denature the protein and the products of phospholipase digestion may inhibit or even activate certain enzymes. Another problem associated with solubilising enzymes is that latent activity may be expressed by the removal of membrane components that block access of substrate to the active site; this cannot be considered as a legitimate activation of the enzyme. The pH conditions under which membrane-bound enzymes operate *in situ* may also differ markedly from the bulk pH in which the activity of a solubilised enzyme is assessed. Thus it is problematic whether changes in enzyme activity are caused *a priori* by damage during the extraction procedure or as a result of subsequent isolation from neighbouring lipid or protein components on which they may rely for their particular functional conformation[16].

1.2.2 Ultrastructural Localisation of Enzymes

A number of enzymes can be localised within the ultrastructure of thin sections by suitable modifications of standard histochemical techniques used in visible-light microscopy[17]. Sections are prepared with fixation treatment sufficient to preserve morphology without substantial inactivation of the enzyme. Accumulation of reaction products alone at the site of formation does not usually provide enough selective enhancement of image contrast so it is often necessary to couple the primary product to a secondary capture reaction involving an electron-dense metal. Final products that yield organic polymers with osmiophilic properties (affinity for osmium stain) are the most sensitive and versatile of these reactions, and the technique can be suitably adapted to detect most hydrolase and oxidase–reductase reactions. Cytochemical studies of this type have enabled enzyme activities to be assigned to precise membrane surfaces and the method is also useful for screening tissues consisting of a number of different cell types. The main disadvantage is the difficulty of preserving morphology while, at the same time, fixing the enzyme within the structure and maintaining its activity. It is also apparent that enzyme activities cannot be expressed in quantitative terms although specific subcellular membranes can be prepared by gentler procedures to provide a correlation with cytochemical localisation.

1.3 Structure and Chemical Composition of Membrane Components

All animal cell membranes consist of an association of lipids and proteins. Different membranes can be characterised broadly on the type and proportion of each of these components. The chemical composition of a particular membrane is not necessarily constant in time but its distinctive identity is usually retained. For example, a protein that is specific for a given membrane does not appear in other morphologically different membranes during the life of the cell; the relative amount of such a protein may vary considerably with time or degree of differentiation of the membrane and in some circumstances it may not be detected at all. These changes may be required to regulate the functional activity of the membrane or they may represent stages in differentiation of the structure. Changes in composition can also be induced by variations in temperature or nutritional status, or by hormone or drug administration.

1.3.1 Membrane-bound Proteins

As we have seen, membranes possess a variety of proteins with demonstrable enzyme activities, but not all membrane proteins are enzymes. Other proteins participate, for example, in immune reactions or act as carrier or transport proteins for the movement of molecules across the membrane, while the function of many other membrane proteins is obscure. The influence of proteins in membrane structure is not yet clear but it is likely that they exert a limited organisational effect on other membrane components in their immediate vicinity.

A number of proteins can usually be removed from membranes by relatively mild extraction procedures, suggesting that they may not be extensively integrated into the structure but simply adsorbed to the surface[18]. This raises the question as to which proteins are truly membrane-bound, in the sense that they form part of the structure or are essential to the function of the membrane, and which proteins are contaminants. Ideally one would include all proteins that associate in a *specific* way with the membrane *in vivo* and exclude those non-specifically adsorbed. In practice, specific associations are defined as those that cannot be broken by repeated washing of the membrane with physiological solutions. Contamination with exogenous adsorbed protein is particularly noticeable in the preparation of red-cell membranes, which adsorb small amounts of haemoglobin, but it is likely that membranes from other sources are also contaminated in this way with less conspicuous proteins. This problem is especially serious in view of the fact that proteins that are normally free in the cytoplasm may adsorb to the membrane non-specifically on manipulation of the ionic conditions and, in the absence of evidence to the contrary, they are likely to be regarded as specific membrane proteins. Another difficulty is that soluble proteins can become trapped inside membrane vesicles created during homogenisation of the tissue and obviously no amount of washing will remove them.

1.3.2 Extraction of Membrane Proteins

A certain amount of information about the nature of the interaction of proteins with other components of the membrane can be obtained from the conditions required to extract them from the membrane. The methods already outlined for extracting enzymes from membranes are generally applicable to other membrane proteins but usually a more systematic approach is adopted when a comprehensive characterisation of these proteins is desired. Some proteins, for example, are released from membranes by raising or lowering the ionic strength or adjusting the pH from neutrality. Procedures employing low ionic strength are usually combined with the addition of chelating agents, and about 25 per cent of the total protein can be extracted from erythrocyte membranes by this method. This protein, which represents three distinct molecular species, is probably bound to the membrane predominantly by coulombic forces, but similar extraction procedures applied to other membranes often release protein complexed with membrane lipids, in which case such conclusions are more tenuous.

Membrane proteins, however, are in general more intractable and require drastic measures to extract them. Such procedures usually violate the structural integrity of the membrane and lead to denaturation of the protein. Partial release can be often achieved with agents like guanidine hydrochloride or urea, which are known to prevent intermolecular hydrogen-bonding and to denature soluble proteins by disrupting secondary and tertiary structure. Organic solvents such as short-chain alcohols, for example butanol or pentanol, as well as aqueous pyridine or acidified phenol are also used but, in common with reagents causing a disruption of hydrogen bonds, there is often an incomplete extraction of protein. In many instances, the extracted protein is combined with membrane lipids, but the latter may be subsequently removed by a further extraction with mildly polar solvents like 90 per cent aqueous acetone or 80 per cent aqueous ethanol.

The most effective method for solubilising membrane protein is undoubtedly the use of surfactant detergents. These agents have a high affinity for membrane constituents and are effective in low concentration. The detergents are classified according to their polar groups; thus nonionic detergents rely on dipole–dipole interactions between hydroxyl groups and water, whereas ionic detergents exhibit strong positive or negative charge–dipole interactions. The structures of three detergents commonly used to solubilise membrane proteins are presented in figure 1.9. The cationic detergents usually possess a quaternary nitrogen functional group such as cetyl trimethylammonium bromide and the anionic

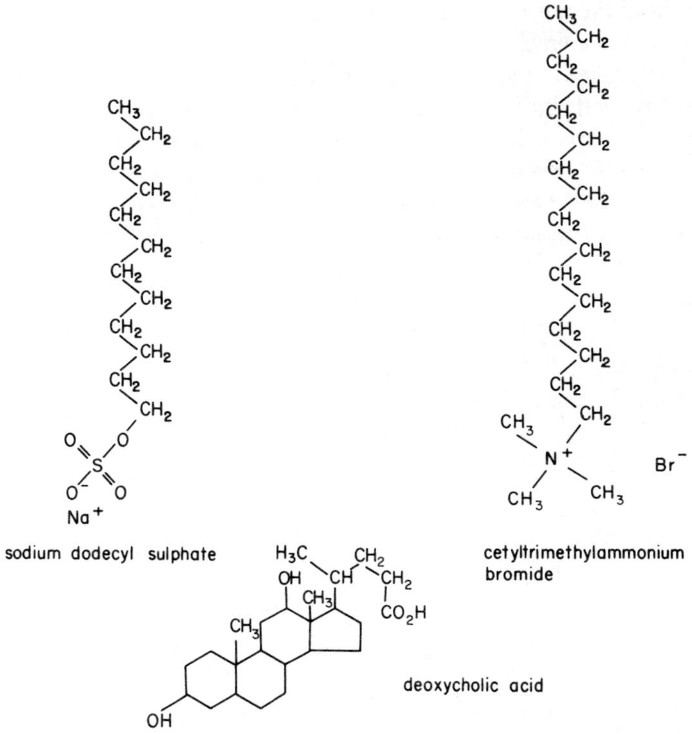

Figure 1.9 Chemical structure of detergents commonly used to solubilise membrane proteins. The surface-active properties of these molecules arise from a polar distribution of a hydrophobic moiety and a hydrophilic ionisable group.

detergents may have carboxyl (bile acids), phosphate (cetyl phosphoric acid) or sulphate (sodium dodecyl sulphate) groups. Sodium dodecyl sulphate possesses special properties in this regard because it dissociates membrane proteins into monomeric subunits and binds to the protein predominantly by hydrophobic interactions in a stoichiometry of 1.4 g of detergent per gram of protein. The interaction confers a regular helical structure on the polypeptide chain, which becomes bent into the shape of a hairpin. The length of the resulting complex is consequently a function of the length of the polypeptide chain and, because of the predictable conformation combined with a constant charge to relative molecular mass ratio, such material is readily amenable to accurate hydrodynamic and electrophoretic studies.

1.3.3 Physical Properties of Membrane Proteins

It is clear from the preceding discussion that most membrane proteins are markedly insoluble in physiological salt solutions. This is because the interaction of membrane proteins with polar solvents is not energetically favourable compared with their interaction with other membrane components, which presumably provide a more hydrophobic environment. On the other hand, proteins such as cytochrome c behave in most respects like typically soluble proteins and this is reflected in the ease with which cytochrome c can be dissociated from the mitochondrial membrane (see section 5.3.6).

Cell membranes contain characteristic proteins, some of which vary in relative proportion according to the functional activity of the membrane. A common method used to distinguish membranes on the basis of their protein content is to determine the relative molecular mass distribution and relative amount of each protein extracted from the membrane[19]. Several methods commonly employed to estimate the relative molecular mass of membrane proteins are illustrated in figure 1.10. Gel-exclusion chromatography is performed by eluting samples of

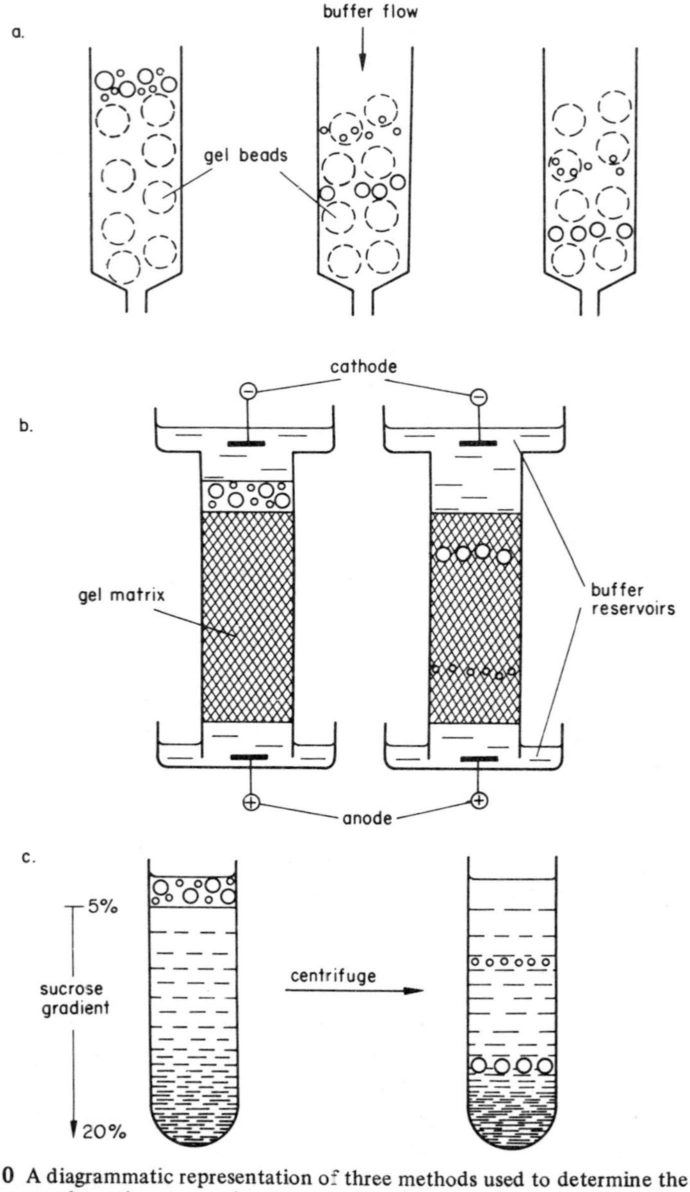

Figure 1.10 A diagrammatic representation of three methods used to determine the relative molecular mass of membrane proteins: (a) gel-exclusion chromatography; (b) polyacrylamide gel electrophoresis; (c) density-gradient sedimentation analysis.

solubilised protein through a column of swollen gel beads – a number of grades of polydextrans (Sephadex) are available – in which two compartments can be recognised; the medium outside the gel of volume V_o and an inner gel matrix of volume V_i. Thus the total bed volume, $V_t = V_o + V_i$. Molecules are separated according to their ability to penetrate the gel matrix. This is related to their size and is expressed as a ratio, $K = (V_e - V_o)/V_i$, where V_e is the volume of the buffer that must pass through the column before the protein peak emerges. Very small molecules will have complete access to V_i and will emerge after a volume V_t has been eluted, whereas molecules completely excluded from the gel pass out in the void volume, V_o. Membrane proteins are generally filtered through columns equilibrated with detergent so that aggregation of the protein can be avoided.

Another procedure is electrophoresis in polyacrylamide gels containing the detergent, sodium dodecyl sulphate (the structure of this compound is shown in figure 1.9). The gel matrix in this system consists of a continuous lattice of covalently cross-linked hydrated polymers. The detergent–polypeptide complex migrates towards the anode when a potential gradient is applied along the column at a rate inversely related to the logarithm of the relative molecular mass. Thus large molecules migrate more slowly than smaller ones at a constant charge to relative molecular mass ratio. Mobility is generally expressed as the ratio of the distance of migration relative to a more mobile marker dye molecule. Two systems are available, one in which the pH of the gel is maintained at a constant value by employing the same buffer in the gel and the reservoirs, and a discontinuous pH system where buffer anions with different mobility are used in the gel and the reservoirs, respectively.

A third method involves the determination of sedimentation characteristics in the ultracentrifuge. Sedimentation coefficients (velocity per unit centrifugal field) of spherical proteins is related to the radius r and density ρ_p by the expression, $s = 2r^2 (\rho_p - \rho_m)/9\eta$, where ρ_m and η refer to the density and viscosity, respectively, of the surrounding medium (all are expressed in c.g.s. units). Sedimentation coefficients of nonspherical proteins can be obtained by applying an appropriate correction for frictional drag through the medium. Coefficients are usually expressed in Svedberg units, S, under specified sedimentation conditions, where $S = s \times 10^{13}$ s. Rates of protein sedimentation can be measured directly by optical methods in solutions of uniform density or on subsequent recovery from linear density gradients after centrifugation. An equilibrium method can also be used in which protein dissolved in CsCl is centrifuged and the CsCl sediments to form a linear density gradient. The protein concentrates in regions of the gradient with buoyant density comparable with that of protein.

A popular method for a complete analysis of membrane peptides involves the extraction of membrane proteins with sodium dodecyl sulphate followed by electrophoresis in polyacrylamide gels containing the same detergent. As indicated above, gel electrophoresis in detergent is essentially a molecular filtration that separates particles on the basis of their ability to penetrate the gel matrix while migrating along an electropotential gradient. The method is sufficiently sensitive to detect as little as 1 or 2 μg of protein after appropriate staining procedures. There is an inverse relationship between mobility through the gel and the logarithm of the relative molecular mass over the range of 10^4 to 10^5 but, even though the linear relationship no longer holds above this range, it is possible, by selecting suitable marker proteins, to obtain reliable molecular mass estimates of polypeptides twice this size. This range embraces most protein monomers so far recovered from membranes although it is true that some of the largest polypeptides found in animal tissues are membrane components. Although most proteins migrate in sodium dodecyl sulphate–polyacrylamide gels according to relative molecular mass, some polypeptides exhibit anomalous mobility characteristics and unless certain precautions are taken, erroneous estimates can be obtained by this method. Such proteins include those with unusual charge or conformation and peptides containing unreduced sulphydryl groups or covalently bound carbohydrate constituents. It is possible in some cases to correct for anomalous migration – for example, by treating proteins with thiol reducing agents to break disulphide bonds before electrophoresis and in other situations by comparing mobility in gels that have varying degrees of cross-linking with the free electrophoretic mobility of the detergent complex.

Membranes are usually found to contain a heterogeneous assortment of proteins differing in both relative molecular mass and electrophoretic mobility and no single protein or group of proteins appears to be common to all membranes. Rough- and smooth-surface endoplasmic reticulum of rat liver, for example, each has about fifteen major peptides that separate by polyacrylamide gel electrophoresis in the presence of sodium dodecyl sulphate; some peptides are common to both membranes, but others are distinct. Of the twelve major proteins observed in the outer mitochondrial membrane some appear to be identical with proteins from the endoplasmic reticulum, but all with the possible exception of one are different from the twenty-three major proteins recovered from the inner mitochondrial membrane.

1.3.4 Proteins of the Erythrocyte Membrane

The proteins of erythrocyte membranes have been studied in considerable detail both in regard to the conditions required to extract particular proteins from the membrane and also the relative molecular-mass distribution of constituent polypeptides. Tanner and Boxer[20] have investigated proteins of the human erythrocyte membrane by selective extraction and characterisation by detergent gel electrophoresis (see figure 1.11). The erythrocyte membrane contains six major protein components comprising approximately two-thirds of the total membrane protein. The remaining protein is distributed among nine other bands visible in sodium dodecyl sulphate–polyacrylamide gels. When washed membrane preparations are exposed to low salt concentration at pH 9.5 in the presence of a chelating agent a high relative molecular-mass protein, spectrin (bands A and B), is extracted from the cytoplasmic surface of the membrane together with another as yet uncharacterised protein (J). Spectrin is known to exhibit anomalous mobility characteristics on sodium dodecyl sulphate-polyacrylamide gel electrophoresis and more reliable relative molecular-mass estimates made by sedimentation equilibrium and viscosity

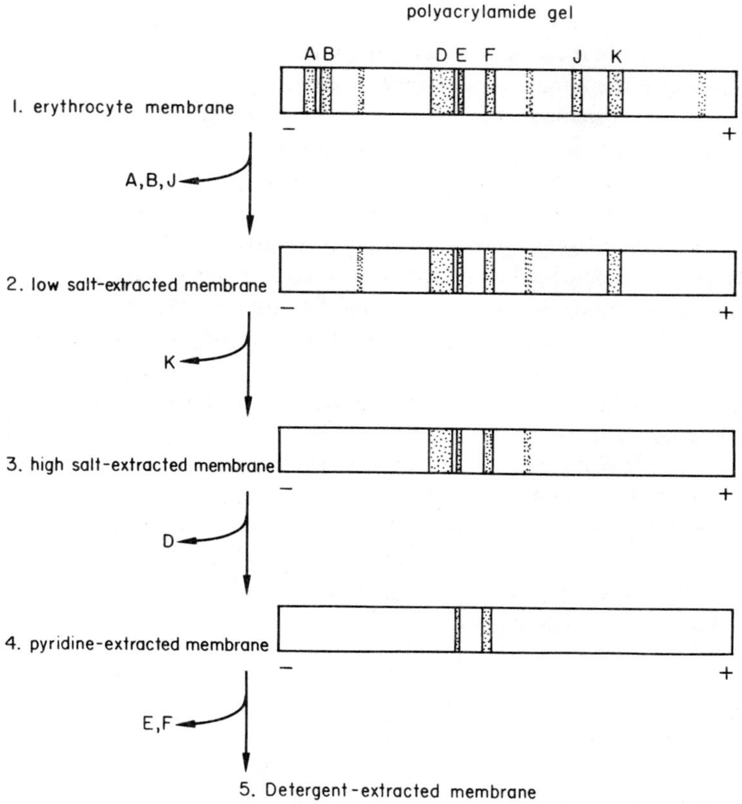

Figure 1.11 Selective extraction of human erythrocyte membrane proteins. Polyacrylamide gel electrophoresis of erythrocyte membrane proteins extracted with sodium dodecyl sulphate (1) show 7 main bands designated A, B, D, E, F, J and K. Treatment of this membrane with low ionic strength buffer in the presence of chelating agent removes bands A, B (spectrin) and protein J (uncharacterised). Subsequent extraction of these membranes (2) with molar sodium chloride removes band K (D-glyceraldehyde 3-phosphate dehydrogenase). Further treatment of salt-extracted membranes (3) with 33 per cent aqueous pyridine removes D (sialoglycoprotein) and only bands E and F remain (4). These can be extracted with sodium dodecyl sulphate. (see reference 20)

measurements place the value close to 140 000. This behaviour is shared by the glycoproteins (see section 1.3.5), which bind proportionately less detergent per unit mass than other proteins and consequently migrate more slowly and at varying rates through gels of differing porosity. Electron microscopy of freeze-cleaved membranes which have been extracted in low salt concentration show the same density of membrane-associated particles as the original membrane preparation, indicating that spectrin and probably also protein J do not penetrate extensively into the interior hydrophobic region of the membrane. This fact, together with the conditions of extraction, suggests that both of these proteins are attached to the membrane predominantly by electrostatic interactions.

Subsequent exposure of the membrane to molar sodium chloride releases almost quantitatively another protein, designated K, which has been identified as the enzyme D-glyceraldehyde 3-phosphate dehydrogenase. As expected, nearly all the enzyme activity originally associated with the membrane can be recovered in the salt extract. High salt concentration also causes a disruption of membrane structure, with the formation of small membrane vesicles in which the remaining components are likely to have undergone considerable rearrangement. Freeze-cleaved fracture faces of the resulting vesicles reveal no membrane-associated particles. At this stage about half of the original membrane protein has been removed. Further extraction of the membrane vesicles with 33 per cent aqueous pyridine removes protein D, which is reputed to carry the M,N antigenic system of the human erythrocyte. Sodium dodecyl sulphate-polyacrylamide gel electrophoresis of this protein suggests a relative molecular mass varying between 65 000 and 95 000, but by using gel filtration the value appears closer to 46 000. The sedimentation characteristics of protein D in the ultracentrifuge, however, imply an even lower and probably more reliable figure of only 24 000. The residual proteins remaining after pyridine extraction, proteins E and F, can be solubilised with detergent and when characterised by gel electrophoresis have relative molecular masses of 93 000 and 74 000 respectively. The large amount of membrane protein that can be extracted from erythrocytes under relatively mild conditions (that is, high or low salt concentrations) is not typical of other subcellular membranes in which a larger proportion of the protein appears to be more securely bound to the membrane.

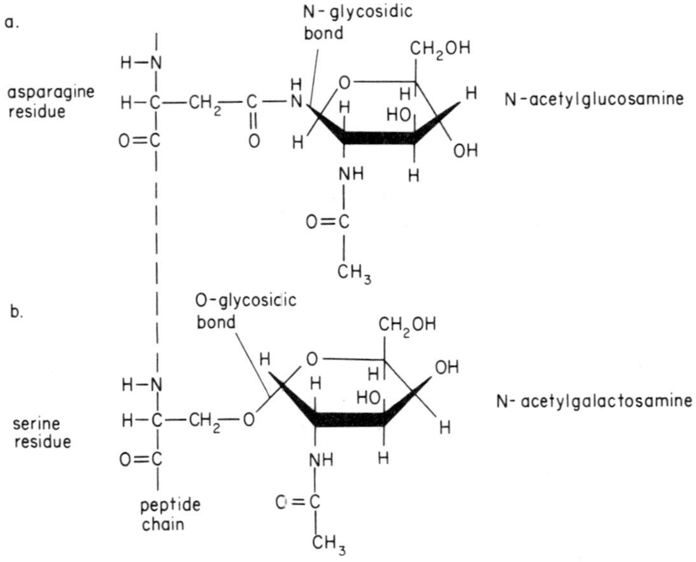

Figure 1.12 Covalent linkages between peptide chains and oligosaccharide in glycoproteins. These are either N-glycosidic bonds between N-acetylglucosamine and asparagine residues (a), or O-glycosidic bonds between N-acetylgalactosamine and serine residues (b).

1.3.5 Chemical Composition of Membrane Proteins

The proteins of plasma membranes can be classified broadly into two groups depending on whether or not they contain carbohydrate in the form of covalently bound neutral and amino sugars[21]. The amount and type of sugar varies from one membrane type to another but most mammalian plasma membranes contain between 2 and 10 per cent by weight of carbohydrate. With the exception of Golgi membranes, most intracellular membranes have very little detectable carbohydrate (see section 3.4.1). The sugars are attached to the polypeptide chain by alkali-labile, O-glycosidic linkages between the side-chain carbon of acetylgalactosamine and hydroxyl groups of serine or threonine. Alternatively, alkali-stable, N-glycosidic bonds are formed between the reducing group of acetylglucosamine and the amide group of asparagine residues (see figure 1.12). The sugars are usually short heterogeneous oligosaccharides ordered in specific characteristic sequences.

Glycoproteins comprise about 10 per cent of the total protein of human erythrocyte membranes and the carbohydrate component constitutes about 60 per cent of the weight of these glycoproteins. Glycoproteins can be distinguished from other polypeptides by the reaction of their constituent sugar residues with periodic acid–Schiff's base. This reaction involves an oxidation by periodic acid of carbohydrate vicinal hydroxyl groups to the corresponding aldehyde and subsequent staining with aldehyde reagents such as acid fuchsin. When sodium dodecyl sulphate–polyacrylamide gels of erythrocyte proteins are stained with periodic acid–Schiff's base, three major glycoproteins can be detected corresponding to proteins D, E and F of figure 1.11. Several other glycoproteins are consistently observed in these gels but they constitute only minor membrane components. It is noteworthy that none of the major glycoproteins can be extracted from the membrane with salt and all are attached through hydrophobic bonding to the membrane. The major glycoproteins have been purified; the amino acid composition and sugar analyses are presented in table 1.2. The amino acid composition of spectrin and D-glyceraldehyde 3-phosphate dehydrogenase from human erythrocyte membrane and cardiac muscle, respectively, are included for comparison.

An inspection of the amino acid composition of these proteins indicates no apparent correlation between the proportion of polar to more hydrophobic amino acids and the conditions required to extract them from the membrane. Indeed by this criterion alone one would expect the glycoproteins to be released most readily under mild conditions on account of the preponderance of polar sugar residues. The explanation of this appears to reside in the distribution of polar and less polar amino acid residues along the polypeptide chain. This concept receives support from amino acid analyses of peptides derived from tryptic digestion of membrane glycoprotein: parts of these are exposed to trypsin while other regions of the polypeptide chain containing less polar amino acids appear to be protected against hydrolysis because of their location within the hydrophobic region of the membrane[24].

Table 1.2 Chemical composition of selectively extracted protein from human erythrocyte membrane

Amino acid	Low ionic strength spectrin (a)	High ionic strength GPDH (b)	Pyridine sialoglycoprotein (c)	Residue E (c)	Residue F (c)
A. lys	6.7	7.6	3.7	5.7	5.2
his	2.6	2.8	5.4	3.2	2.5
arg	5.8	2.8	5.3	5.7	6.1
glu	20.5	6.7	9.7	10.4	11.1
asp	10.9	12.5	5.8	9.9	8.7
total	47.5	32.4	29.9	34.9	33.6
B. gly	4.9	10.5	4.8	9.2	8.1
ala	9.2	9.7	5.1	10.7	9.2
ser	4.1	6.1	14.6	5.2	5.6
pro	2.4	3.8	7.5	5.0	5.4
thr	3.6	6.3	11.5	5.8	5.8
$\frac{1}{2}$cys	1.1	–	0	1.3	2.0
C. val	4.7	8.6	7.7	7.4	7.2
leu	12.4	5.9	5.9	10.8	11.7
ilu	4.0	5.9	7.4	4.3	4.3
tyr	2.0	2.8	2.5	1.4	1.5
phe	3.0	4.1	1.9	4.1	3.7
met	1.7	2.8	1.2	2.3	1.8
try	–	1.2	–	–	–
total	27.8	31.3	26.6	30.3	31.2
Sugar					
fucose			18	7	?
mannose			24	8	6
galactose			135	8	9
glucose			12	200?	32
GlcNAC			50	17	21
GalNAC			118	–	–
sialic acid			135	–	4
rel. mol. mass	140 000	120 000	30 000–50 000	93 000	74 000

The amino acids are subdivided into three groups. Group A consists of the hydrophilic amino acids some of which are basic (lys, his, arg) and others acidic (glu, asp). Group B contains neutral amino acids (gly, ala, ser, pro, thr, $\frac{1}{2}$cys) and Group C, the hydrophobic amino acids (val, leu, ilu, tyr, phe, met, try). The sum (in moles per cent) of amino acids of Groups A and C are shown at the foot of the respective column. Spectrin and glyceraldehyde 3-phosphate dehydrogenase (GPDH) are not glycoproteins and contain no detectable carbohydrate.

(a) Data from Marchesi *et al.*[22]

(b) Heart muscle enzyme, from Allison and Kaplan[23]

(c) Data from Tanner and Boxer[20]

1.3.6 Peptide Mapping and Terminal Amino Acid Analysis

When the complete amino acid composition of a membrane protein has been determined, a procedure known as peptide mapping can be used to check the homogeneity and size of the polypeptide[25]. The protein is hydrolysed by trypsin and the resulting peptides are separated by chromatographic and electrophoretic techniques. The number of peptides released by tryptic digestion ideally equals $n + 1$ where n equals the number of lysine plus arginine residues. Peptide maps of heterogeneous polypeptide mixtures give more peptides than are indicated by examination of the amino acid analysis and correspondingly less of each peptide than expected for an equivalent amount of starting material. Some caution must be exercised when interpreting peptide maps in this way, since a number of membrane proteins have been shown to yield homologous peptides on tryptic digestion. Occasionally bonds resistant to tryptic hydrolysis are encountered and these are generally associated with blocked N-terminal amino groups, in which case a peptide may be released but cannot be detected in the usual manner. Hydrolysis is also inhibited by detergents such as sodium dodecyl sulphate and these must be removed by dialysis or organic-solvent extraction before digestion. The protein may precipitate on removal of detergent but can be solubilised by forming carboxymethyl or sulphonate derivatives, a procedure that does not affect subsequent digestion with trypsin.

Another method of identifying membrane proteins is by end-group analysis where C and N-terminal amino acids are reacted covalently with specific reagents. The amino acid derivatives can then be identified after separation by conventional chromatographic procedures. It is theoretically possible to determine the number of different proteins and also the relative proportions of each. The usefulness of terminal amino acid analyses is, however, limited in practice because many membrane proteins have unreactive or blocked end groups. This is particularly true of the N-terminus and usually only the minimum number of polypeptides can be obtained with any degree of confidence. The reagents most commonly used are dinitrofluorobenzene and 1-dimethylaminonaphthalene-5-sulphonyl chloride (dansyl chloride) for N-terminal reaction and hydrazinolysis, combined with carboxypeptidase A and B digestion, for C-terminal analysis; all react in the presence of sodium dodecyl sulphate. Corrections must be applied to account for any free amino acids that may bind to the protein, since they also react and can be confused with the authentic terminal groups.

1.3.7 Characterisation of Membrane Lipids

The hydrophobic character of cell membranes is almost entirely due to the lipid component and most of these constituents, in monomeric form, are insoluble in water. Virtually complete extraction of membrane lipids can be achieved by treatment with certain nonpolar organic solvents such as chloroform, ether or benzene but a mixture of chloroform and methanol is probably the most effective solvent. Solvents such as acetone are also useful particularly in preparations where only a partial extraction of lipids is desired. Most membrane proteins are denatured by organic solvents and the resulting precipitate can be conveniently removed virtually free of any lipid contamination. Earlier analyses of membrane lipids were based on partial-extraction procedures similar in principle to those now adopted for proteins. Crude lipid fractions were classified on the basis of the method used for extraction from the membrane. Cephalin is a lipid fraction based on selective methods, but it is now known that this extract contains a whole range of distinct phospholipid species.

The major advances in lipid chemistry have been partly due to improved chromatography techniques capable of separating complete lipid extracts into lipid classes as well as resolving the chemical components within each of these classes. Thin-layer or column-chromatographic methods are commonly used to separate the major lipid classes and numerous solvent systems can be combined with support phases of silicic acid, aluminium oxide or cellulose[26]. Susceptibility of certain bonds within the molecule to hydrolysis by acid or alkali provides a further means of distinguishing lipids within each of the main classes. Fatty acyl groups, for example, are hydrolysed in 0.03 N sodium hydroxide whereas fatty ethyl groups are stable to mild alkaline hydrolysis but can be cleaved under acid conditions. Finally, the use of gas–liquid chromatography has proved a most sensitive and reproducible means of completely characterising individual molecular species of lipid. This method consists of degrading complex lipids into constituent moieties (for example, fatty acids, sugars) and preparing apolar volatile derivatives, which can then be partitioned between a hot gas flowing over an apolar liquid phase coated onto an inert support. Detector devices sensitive to a few nanograms of most derivatives record the compounds emerging from the column. The analysis of column effluents by mass spectrometry has introduced an added sophistication to the technique. The use of such sensitive assay methods has emphasised the need to ensure that lipids are not altered chemically during preparation for analysis. These changes can result from *post mortem* enzyme degradation of lipids and are best avoided either by immediate treatment with protein denaturing agents or by maintaining the tissue at low temperature. Oxidation of fatty acyl groups can often be a serious problem but using antioxidants such as butylated hydroxytoluene or preparing and handling lipid extracts under anoxic conditions is helpful in preventing oxidations.

1.3.8 Chemical Composition of Membrane Lipids

All membrane lipid molecules possess regions that are distinctly hydrophobic in character, while others have a greater affinity for a more polar environment. The polar or hydrophilic groups vary in extent and relative affinity for water, ranging from single hydroxyl groups, as in cholesterol, through the zwitterionic and charged groups of the glycerophosphatides to complex oligosaccharides of the glycosphingolipids. This common dichotomy of structural affinity or amphipathicity is a unique feature of membrane lipids and serves to facilitate their mutual interaction and orientation within the structure (see chapter 2).

The lipid components are classified on the basis of chemical structure and in eukaryotic cell membranes they fall principally into four groups; phospholipids, sphingolipids, glycolipids and sterols (see tables 1.3, 1.4 and figure 1.13). Normally all these lipids are confined to cell membranes and, with a few exceptions like lipid-storage organelles and yolk, they are not found free in the cytoplasm in appreciable amounts. A special class of plasma proteins, the plasma lipoproteins, do have affinity for some of these lipids and constitute an important transport system for lipids in the blood.

Table 1.3 The fatty acyl residues (R) commonly found in membrane lipids

Carbon atoms	Fatty Acyl Substituents (R) Chemical structure	Common name of acid
12	$CH_3-(CH_2)_{10}-\underset{\underset{O}{\Vert}}{C}-O-$	lauric
14	$CH_3-(CH_2)_{12}-\underset{\underset{O}{\Vert}}{C}-O-$	myristic
16	$CH_3-(CH_2)_{14}-\underset{\underset{O}{\Vert}}{C}-O-$	palmitic
16	$CH_3-(CH_2)_5-CH{=}CH-(CH_2)_7-\underset{\underset{O}{\Vert}}{C}-O-$	palmitoleic
18	$CH_3-(CH_2)_{16}-\underset{\underset{O}{\Vert}}{C}-O-$	stearic
18	$CH_3-(CH_2)_7-CH{=}CH-(CH_2)_7-\underset{\underset{O}{\Vert}}{C}-O-$	oleic
18	$CH_3-(CH_2)_4-CH{=}CH-CH_2-CH{=}CH-(CH_2)_7-$ $-\underset{\underset{O}{\Vert}}{C}-O-$	linoleic
18	$CH_3-CH_2-CH{=}CH-CH_2-CH{=}CH-CH_2-CH{=}$ $=CH-(CH_2)_7-\underset{\underset{O}{\Vert}}{C}-O-$	linolenic

The glycerophosphatides are the most ubiquitous class of membrane lipids[27]. The basic structure consists of a diglyceride with phosphoric acid esterified to the primary hydroxyl group of the glycerol residue; usually the phosphate is combined in diester linkage with a short base or *myo*inositol. The fatty acyl groups on the 1 and 2 carbon atoms of the glycerol are long unbranched chains varying in length by two carbon atoms and in the extent of saturation. Double bonds located between specific carbon atoms are all in the *cis* configuration. The fatty acyl residue in the 1-carbon position is generally more saturated than the residue located on the 2-carbon position of the glycerol. A separate class of lysoglycerophosphatides contains only one fatty acyl residue, thus providing them with strong detergent-like properties. However, lysoglycerophosphatides are only found occasionally in cell membranes and then only in minor amounts. The

a. Phospholipids

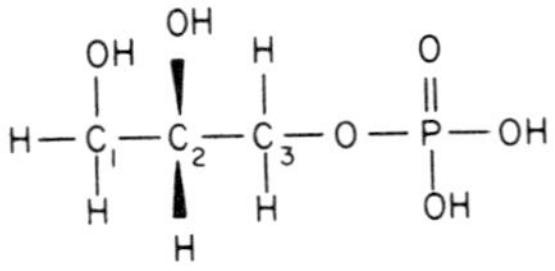

sn-glycerol 3-phosphoric acid

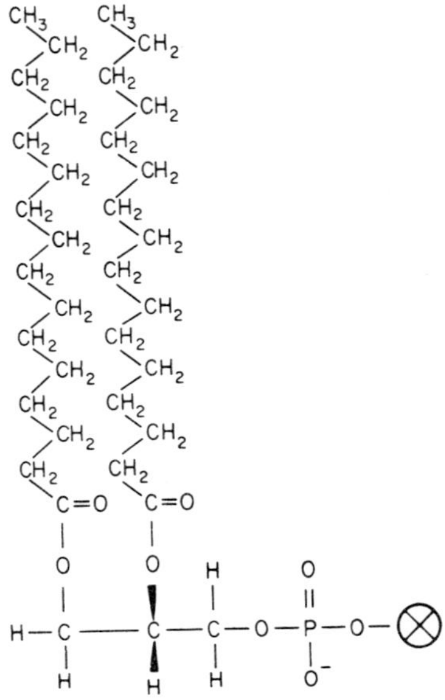

dipalmitoylphosphatides

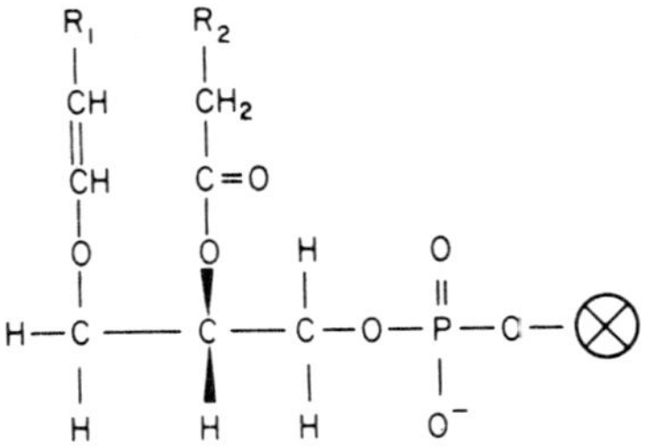

plasmalogens

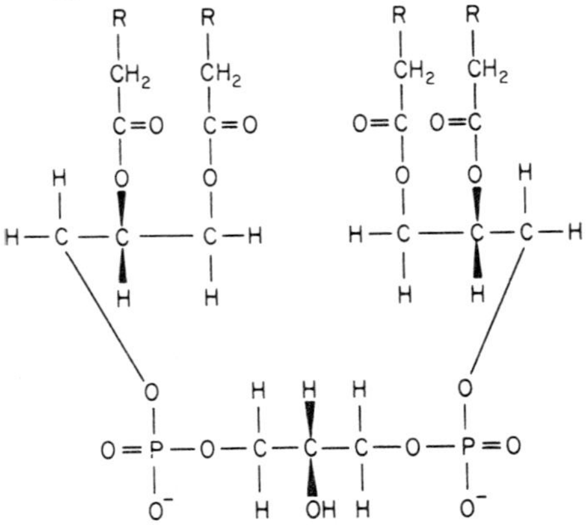

cardiolipin
(polyglycerol
phospholipid)

b. Sphingolipids

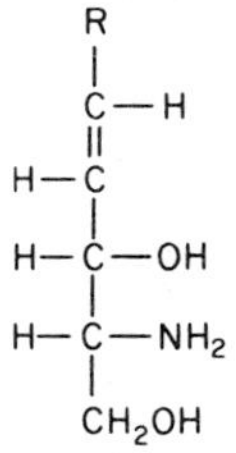

sphingosine

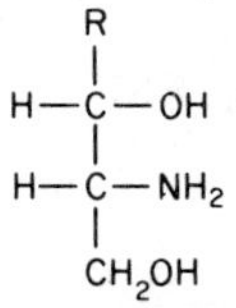

dihydrosphingosine

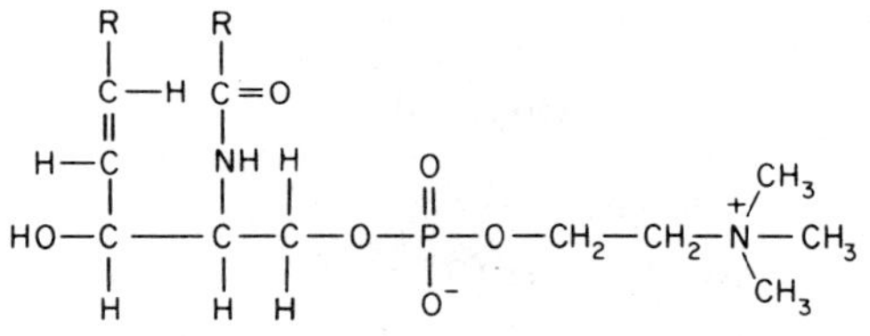

sphingomyelin

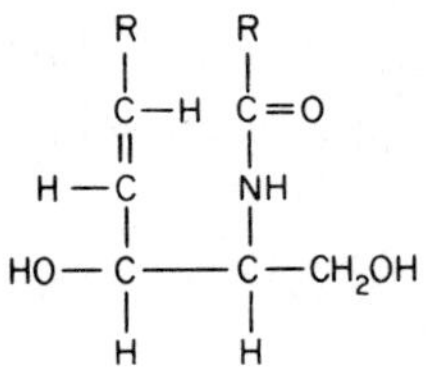

ceramide

c. Glycolipids

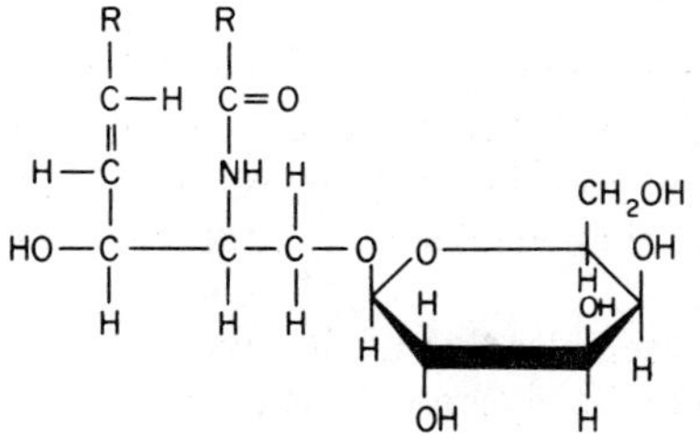

cerebroside

d. Sterols

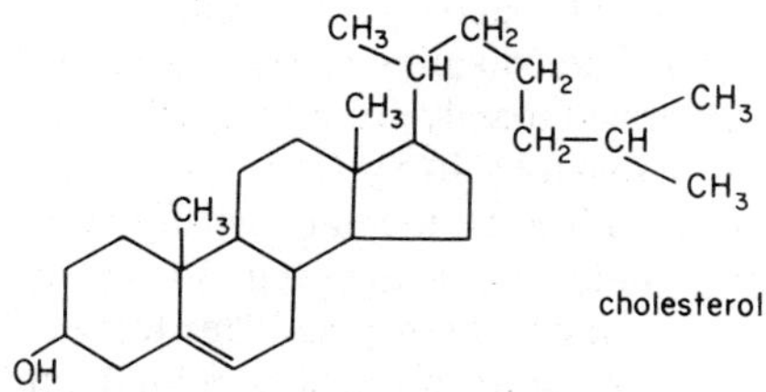

cholesterol

Figure 1.13 Chemical structure of membrane lipids.

Table 1.4 Phosphate ester substituents of membrane phospholipids

Phosphoglyceride	Chemical structure ⊗
Phosphatidic acid	$-H$
Phosphatidylcholine	$-CH_2-CH_2-N^+(CH_3)_3$
phosphatidylethanolamine	$-CH_2-CH_2-NH_2$
phosphatidylserine	$-CH_2-CH(NH_2)-C(=O)-OH$
phosphatidylglycerol	$-CH_2-CH(OH)-CH_2-OH$
phosphatidylinositol	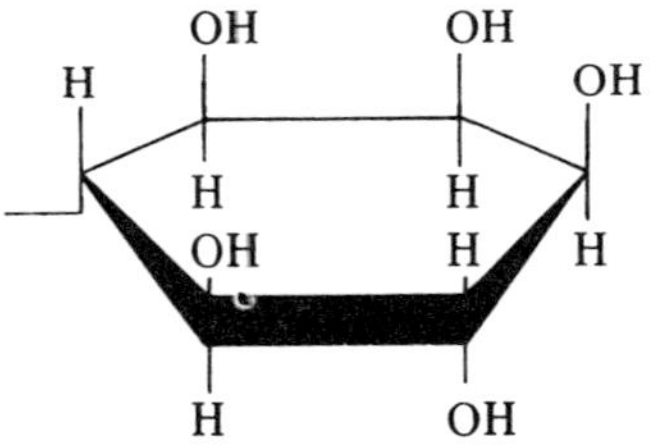

other class of glycerophosphatides is distinguished by an alkali-stable, ether-linked, paraffinic group usually on the 1-carbon position of the glycerol. These are known as plasmalogens and on acid hydrolysis they yield equimolar amounts of long-chain fatty acids and fatty aldehydes.

The structure of the sphingolipids resembles that of the glycerolipids except that a long-chain amino alcohol replaces the glycerol in the pivotal position. Sphingomyelin has an additional fatty acid amide bonded to the amino group of sphingosine as well as a choline in phosphodiester linkage to the terminal hydroxyl group. The ceramides constitute a class of sphingolipids that has a free hydroxyl group in the C-1 position. Esterification of sugars through this hydroxyl group gives rise to the separate class of glycolipids: these can be either simple in structure, possessing only a single galactosyl residue (cerebrosides), or complex oligosaccharides containing neutral and charged amino sugars (gangliosides). The fatty acyl residues of these lipids are derived mainly from long-chain tetracosanoic and tetracosenoic acids and many are stable α-hydroxy fatty acids. Neutral lipids are present in varying amounts in cell membranes; they are usually more abundant in the plasma membrane. Unesterified cholesterol is the predominant neutral lipid of membranes and neutral glycerides, cholesterol esters and free fatty acids are only occasionally observed as minor components of membrane lipid extracts.

1.3.9 Lipid Composition of Cell Membranes

Given that due care is exercised in preparing cell membranes for lipid analyses, the major lipid classes represented in particular membranes are remarkably constant and the proportions do not appear to alter significantly with changing diet or functional activity of the cell. The proportions of protein to lipid and cholesterol to polar lipids in the various membranes of rat liver are presented in table 1.5.

Table 1.5 Protein and lipid ratios of rat tissue membranes

Membrane	Protein/lipid	Cholesterol/polar lipid
plasma membranes		
myelin	0.25*	0.95
erythrocyte	1.1	1.0
rat liver cells	1.5	0.5
plasma membrane		
nuclear membranes	2.0	0.11
endoplasmic reticulum		
rough-surface	2.5	0.10
smooth-surface	2.1	0.11
mitochondrial membranes		
inner membrane	3.6	0.02
outer membrane	1.2	0.04
Golgi membranes	2.4	–

*Myelin from central nervous system; higher ratios are found in peripheral nerve myelin

Corresponding values for myelin and erythrocyte membranes are included for comparison. The relative size of each phospholipid class in these membranes is given in table 1.6. Morphologically distinct membranes from one species do not necessarily have the same lipid composition as similar membranes from another, even though they may perform essentially the same function in both species. This point is illustrated in figure 1.14a, where it can be seen that, of the major erythrocyte lipids, sphingomyelin together with phosphatidylcholine account for 40 to 45 per cent of the total membrane lipid from rat, pig, ox and sheep, but the ratio between the two phospholipids varies widely in the different species. That these differences are not confined to membranes from different species is exemplified by the lipid composition of membranes of the endoplasmic reticulum (a mixture of smooth- and rough-surface membranes) from different bovine tissues (see figure 1.14b). Sphingomyelin and phosphatidylcholine in this case represent between 50 and 60 per cent of the total phospholipid extracted from purified membranes from kidney, heart and liver, yet the ratio of sphingomyelin to phosphatidylcholine again varies widely.

Although the main lipid classes appear to be invariant for a particular membrane, this does not usually apply to the substituent molecular species within each of the major lipid groups. The most notable differences arise in the fatty acyl substituents, particularly of the phospholipids, where length and extent of saturation may reflect adaptation to the prevailing physical requirements of the membrane[29].

Subcellular membranes all contain phospholipids but a few common distinctive lipids are present in each membrane type. The plasma membrane, for example,

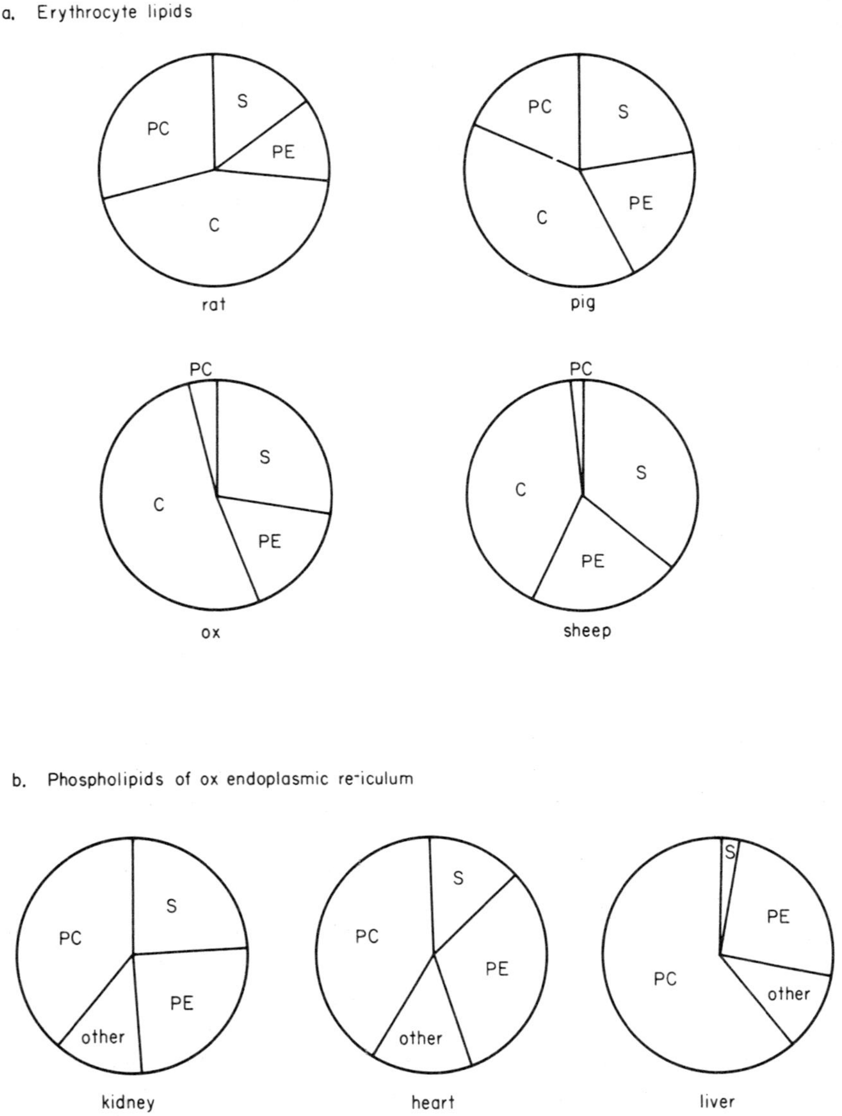

Figure 1.14

(a) Variation in the proportion of sphingomyelin (S) to phosphatidylcholine (PC) in the total lipid fraction of erythrocytes from various species. The remaining lipids consist mainly of cholesterol (C) and phosphatidylethanolamine (PE). The sum of sphingomyelin plus phosphatidylcholine constitutes a relatively constant proportion of the erythrocyte membrane lipids, but the ratio between the two phospholipids varies considerably in different species.

(b) Sphingomyelin (S) and phosphatidylcholine (PC) relative to total phospholipids present in the endoplasmic reticulum (smooth-surface plus rough-surface membranes) of different ox tissues. (phosphatidylethanolamine = PE).

Table 1.6 Phospholipid composition of liver cell membranes

Phospholipid	Plasma membrane	Nuclear membranes	Rough endoplasmic reticulum	Golgi membranes	Mitochondrial membranes		Lysosomal membranes
					inner	outer	
phosphatidylcholine	34.9	61.4	60.9	45.3	45.4	49.7	33.5
phosphatidylethanolamine	18.5	22.7	18.6	17.0	25.3	23.2	17.9
phosphatidylinositol	7.3	8.6	8.9	8.7	5.9	12.6	8.9
phosphatidylserine	9.0	3.6	3.3	4.2	0.9	2.2	8.9
phosphatidylglycerol	4.8	*	*	*	2.1	2.5	*
phosphatidic acid	4.4	1.0	1.0	*	0.7	1.3	6.8
cardiolipin	trace	0	*	*	17.4	3.4	6.8
lyso phosphatidylcholine	3.3	1.5	4.7	5.9	*	*	0
lyso phosphatidylethanolamine	*	0	0	6.3	*	*	*
sphingomyelin	17.7	3.2	3.7	12.3	2.5	5.0	32.9

*no value available. All values expressed as moles per cent of total phospholipid.

Collated by McMurray and McGee[28]

contains most of the cellular glycolipids; these lipids are not usually found in mitochondria, endoplasmic reticula or nuclear membranes. The plasma membrane also contains a high proportion of the total cell membrane cholesterol. Myelin, which we have seen is a special elaboration of the plasma membrane (section 1.1.1), is particularly rich in glycolipids and contains most of the cerebroside and the corresponding sulphate ester of cerebroside (cerebroside containing a sulphate group attached to the 3-position of the galactosyl moiety) found in the body. Nuclear membranes have a characteristic phospholipid complement, which distinguishes them from other subcellular membranes. About half of the total phospholipid of this membrane is in the form of phosphatidylcholine and together with phosphatidylethanolamine and phosphatidylinositol accounts for more than 75 per cent of the membrane phospholipid. Nuclear membranes from different tissues or even between different vertebrate species appear to have similar phospholipid compositions. The phospholipid content of rough- and smooth-surface endoplasmic reticula is similar and this is generally the case for these membranes in the same organ of different species although variations have been observed between different organs of the same species. Membranes of the Golgi complex have a phospholipid composition intermediate between endoplasmic reticula and the plasma membrane and this is considered to be evidence supporting the concept that Golgi membranes form transition membranes between them. Mitochondrial membranes are characterised by the presence of cardiolipin which is confined almost entirely to this organelle and is located predominantly in the inner membrane. Sphingolipids are found in only small amounts and cholesterol is distributed mainly in the outer mitochondrial membrane. The lipid composition of mitochondria does not vary significantly between tissues of the same species and there are only slight variations between different vertebrate species. The means whereby particular membranes achieve and retain their distinctive lipid composition is one of the most perplexing and, as yet, unresolved problems of membrane biochemistry.

1.3.10 Pathological Disturbances of Membrane Lipid Composition

Two types of membrane lipid abnormalities are recognised and both can be attributed to hereditary sources[30]. The first group includes cystic fibrosis of the pancreas, renal glycosuria, cystinuria and renal tubular acidosis and the Fanconi syndrome among others. It should be noted that a defect in the metabolism of any particular membrane lipid component can often lead to a disturbance of other lipid constituents such that the primary effect may be disguised by the secondary responsive changes. Nevertheless, all these conditions are believed to be true primary defects of membrane lipid metabolism although the precise mechanisms, in molecular terms, have not been established for any of them.

The second group involves defects in the enzymes concerned with sphingolipid metabolism some of which are confined mainly to the nervous system while others have repercussions in a wider sphere of tissues. The disorders are characterised by an accumulation of specific membrane sphingolipids resulting from failure to co-ordinate synthesis and degradation of the lipid; the defect usually involves the inability to cause breakdown of the lipid. The original concept that each sphingolipidosis arises as a consequence of a single gene mutation has given way to a broader classification in the light of clinical, genetic and biochemical evidence suggesting that each condition is a distinct pathological state. Because more detail of the lipid chemistry is now known, they are classified according to the lipid concerned. The specific disorders include sphingomyelinosis (Niemann–Pick disease), gangliosidosis (Tay–Sachs disease, including the juvenile form) and cerebrosidosis (Gaucher's disease). Disturbances in cerebroside sulphate (Metachromatic Leucodystrophy) and ceramide polyhexoside metabolism (Fabry's disease) have also been documented. No hereditary disturbances involving phosphoglyceride metabolism have been described in which a specific phospholipid accumulates in membranes or cells. There are, however, some instances where the metabolic turnover of these lipids is out of balance with the specific functional requirements of a particular membrane. An example of this situation is where the level of unsaturated bonds of the fatty acyl residues cannot be maintained in the face of rapid autoxidation and the oxidised lipid accumulates in the form of lipofucin deposits called ceroid.

1.4 Summary

Cell membranes can be characterised by three methods: 1. ultrastructural associations; 2. enzyme specificity; 3. chemical composition.

Most subcellular membranes of chemically fixed and stained tissue sections, when viewed under the electron microscope, appear as trilamellar structures. The electron-density profile of these structures depends on where the stain, used to develop contrast in the specimen, is deposited and does not give a reliable indication of the underlying molecular arrangement. Particular cell membranes are identified either by their position within the cell – for example, the plasma membrane forms the cell boundary while the nuclear envelope surrounds the nucleus – or by morphology of the organelles from which they are formed such as mitochondria, endoplasmic reticulum or the Golgi complex. The development of physical fixation techniques involving rapid freezing and subsequent examination of the cleaved fracture profiles by electron microscopy may provide a better method of identifying membranes based on the molecular detail revealed in particular membranes.

Membranes possess a variety of different enzyme activities and these enzymes may be more or less an integral part of the structure. Certain enzymes are found exclusively in specific membrane types, providing another means by which they can be identified. Membrane-bound hydrolase and oxidase–reductase enzymes can be localised within the ultrastructure by cytochemical techniques and in some cases localisation to precise membrane surfaces is possible. Enzyme activites can be used to check membrane purity during isolation procedures and identify the source of contaminating membranes.

Cell membranes are composed of a characteristic array of various lipids and proteins. Some flexibility is apparent in the type and relative amount of protein in membranes, which is related to the degree of membrane differentiation or in some cases to functional activity. Some membrane proteins contain carbohydrate in the form of oligosaccharides and these glycoproteins are confined almost entirely to the plasma membrane. The association of protein with the membrane varies from weak specific interactions, which can be broken by manipulation of the electrostatic conditions, to strong interactions requiring detergent extraction. There is no apparent correlation between the amino acid composition of intact membrane proteins and the conditions required to extract them from the membrane to indicate that their composition adapts them to this particular function.

Membrane lipids are all amphipathic and they can be conveniently extracted with organic solvents. Four lipid classes are represented in eukaryotic cell membranes:

1. Glycerophosphatides, which are found in all cell membranes and consist of a glycerol moiety to which long-chain fatty acids are acylated. The phosphate group is usually in diester linkage with a short-chain base or *myo*inositol.
2. Sphingolipids with a similar structure to the glycerophosphatides but with the base, sphingosine, replacing glycerol.
3. Glycolipids consisting of ceramide to which single hexoses or complex oligosaccharides are esterified through the hydroxyl group at the C-1 position.
4. Sterols, of which cholesterol is the most notable representative.

The combination of lipid classes found in particular membranes is relatively constant but there is a considerable variation in substituent fatty-acid chain length and degree of unsaturation within appropriate lipid classes. Hereditary defects in sphingolipid metabolism have been well documented in man and these conditions can give rise to secondary disturbances in membrane-lipid composition.

References

1. D. W. Fawcett, *The Cell, an Atlas of Fine Structure,* Saunders, Philadelphia (1966)
2. K. R. Porter and M. A. Bonneville, *Fine Structure of Cells and Tissues,* Kimpton, London (1963)
3. C. M. Feldherr. Structure and function of the nuclear envelope. *Adv. cell. molec. Biol.,* **2** (1972), 273–307
4. J. Markovics, L. Glass and G. G. Maul. Pore patterns on nuclear membranes. *Expl Cell. Res.,* **85** (1974), 443-51
5. P. J. Goldblatt. The endoplasmic reticulum. In: *Handbook of Molecular Cytology,* (A. Lima de Faria ed.), North Holland, Amsterdam (1969), pp 1101–29
6. P. Favard. The Golgi apparatus. In: *Handbook of Molecular Cytology,* (A. Lima de Faria ed.), North Holland, Amsterdam (1969), pp 1130–55
7. E. A. Munn, *The Structure of Mitochondria,* Academic Press, London (1974)
8. F. S. Sjostrand, *Electron Microscopy of Cells and Tissues,* Academic Press, London (1967)

9. H. T. Meryman. Mechanics of freezing in living cells and tissues. *Science, N.Y.*, **124** (1956), 515-21
10. D. W. Deamer and D. Branton. Fracture planes in an ice-bilayer model membrane system. *Science*, N.Y., **158** (1967), 655-7
11. D. Branton. Freeze-etching studies of membrane structure. *Phil. Trans. R. Soc., Lond. B.*, **261** (1971), 133-8
12. K. Muhlethaler. Studies on freeze-etching of cell membranes. *Int. Rev. Cytol.* **31** (1971), 1-19
13. H. D. Brown and S. K. Chattopadhyay. Organelle-bound enzymes. In: *Chemistry of the Cell Interface*, (H. D. Brown ed.), Academic Press, New York (1971), pp 73-203
14. J. N. de Pierre and M. L. Karnovsky. Plasma membranes of mammalian cells. A review of the methods for their characterization and isolation. *J. Cell Biol.*, **56** (1973), 275-303
15. D. H. De Heer, M. S. Olsen and R. N. Pinckard. Characterization of rat liver subcellular membranes. Demonstration of membrane-specific autoantigens. *J. Cell Biol.*, **60** (1974), 460-72
16. R. Coleman. Membrane bound enzymes and membrane ultrastructure. *Biochim. Biophys. Acta*, **300** (1973), 1-30
17. T. K. Shnitka and A. M. Seligman. Ultrastructural localization of enzymes. *A. Rev. Biochem.*, **40** (1971), 375-96.
18. G. Guidotti. Membrane proteins. *A. Rev. Biochem.*, **41** (1972), 731-52
19. T. L. Steck and C. F. Fox. Membrane proteins. In: *Membrane Molecular Biology*, (C. F. Fox and A. D. Keith eds), Sinauer Associates, Stamford, Conn. (1972), pp 27-75
20. M. J. A. Tanner and D. H. Boxer. Separation and some properties of the major proteins of human erythrocyte membrane. *Biochem. J.*, **129** (1972), 333-47
21. G. A. Jamieson and T. J. Greenwalt; (eds), *Glycoproteins of Blood Cells and Plasma*, Lippincott, Philadelphia (1971)
22. S. L. Marchesi, E. Steers, V. T. Marchesi and T. W. Tillack. Physical and chemical properties of a protein isolated from red cell membranes. *Biochemistry*, **9** (1970), 50-7
23. W. S. Allison and N. O. Kaplan. The Comparative enzymology of triosephosphate dehydrogenase. *J. biol. Chem.*, **239** (1964), 2140-52
24. V. T. Marchesi, T. W. Tillack, R. L. Jackson, J. P. Segrest and R. E. Scott. Chemical characterization and surface orientation of the major glycoprotein of the human erythrocyte membrane. *Proc. natn. Acad. Sci. U.S.A.*, **69** (1972), 1445-9
25. D. M. Kaplan and R. S. Criddle. Membrane structural proteins. *Physiol. Rev.*, **51** (1971), 249-72
26. G. V. Marinetti (ed.) *Lipid Chromatographic Analysis*, Vol. I, Dekker, New York (1967)

27. G. B. Ansell and J. N. Hawthorne, *Phospholipids, Chemistry, Metabolism and Function.* Elsevier, Amsterdam (1964)

28. W. C. McMurray and W. L. Magee. Phospholipid metabolism. *A. Rev. Biochem.*, **41** (1972), 129–60

29. G. Rouser, G. J. Nelson, S. Fleischer and G. Simon. Lipid composition of animal cell membranes, organelles and organs. In: *Biological Membranes, Physical Fact and Function,* (D. Chapman ed.) Academic Press, London (1968), pp 5–69

30. R. O. Brady. Disorders of lipid metabolism. In: *Current Trends in the Biochemistry of Lipids,* (J. Ganguly and R. M. S. Smellie eds), Academic Press, London (1972), pp 113–27

2 Physical Structure of Membrane Components

The chemistry of membrane constituents provides almost unlimited scope for variety, particularly with regard to the hydrocarbon residues of the various lipid classes. It might be anticipated from the complexities involved that any attempt to interpret membrane structure and function on the basis of chemical composition alone would prove an extremely difficult task. Nevertheless, the physical properties of lipids and proteins is largely a function of the disposition of chemical groups within each molecule, some of which are hydrophobic, some hydrophilic, while others are more or less accommodated in either environment. All membrane lipids, without exception, are amphipathic since most strongly exhibit both hydrophobic and hydrophilic affinities. The reason for this can be found in the chemical nature of the respective groups and the fact that hydrophobic residues are discretely separated from the polar groups within the molecule. Amphipathicity of membrane proteins is considerably more difficult to establish because this property ultimately depends on the disposition of a large number of residues which, individually, react only weakly to polar or nonpolar environments. The solubility characteristics of membrane proteins in solvents of differing polarity and their propensity to interact with hydrocarbon residues of membrane lipids and other detergents tends to suggest that these proteins also possess amphipathic properties.

Because there are fundamental differences in the character of lipids and proteins, it has been necessary to develop separate techniques to assess their physical characteristics. Membrane lipids, for example, form characteristic structures when dispersed in water and the physical properties of these structures resemble in many respects those of biological membranes. The obvious advantage of such systems is that purified lipid fractions or synthetic lipids of defined molecular structure can be examined under a variety of conditions enabling individual components of a membrane lipid extract to be studied in isolation or in selected combinations to build up an overall picture of their contribution to membrane structure. In general, membrane proteins cannot be separated from other membrane constituents without compromising their structure and for this reason studies of membrane-protein conformation are usually restricted to determinations of net protein conformation of intact membrane preparations. Conventional methods for assessing protein structure have been developed mainly for pure water-soluble proteins, and studies of protein mixtures in particulate suspension present a number of problems. Among the most serious of these is the fact that phyical parameters obtained from a mixture of insoluble membrane proteins must be interpreted from the behaviour of model amino acid polymers, which assume well-defined conformations in aqueous solution. Extrapolation of such information may be regarded as rather tenuous particularly since the microenvironment of the membrane is likely to dictate much of the conformational arrangement of these proteins. Another problem, common to physical measurements of proteins as well as lipids, is that these measurements reflect only an average situation pertaining to an assembly of molecules and consequently the properties of individual components in diverse mixtures cannot be readily evaluated. In studies of protein conformation in membranes, for example, each protein contributes to the average protein conformation to an extent depending on its relative concentration in the mixture.

2.1 Protein Structure

2.1.1 General Concepts in Membrane Protein Conformation

Studies of the structure of amino acid polymers by X-ray diffraction and other methods has shown that the secondary and tertiary structure of polypeptides is a function of amino acid composition and the disposition of these residues along the polypeptide chain. The chain becomes folded to accommodate the net forces acting between the various side chains and the surrounding solvent molecules, and between other amino acids in either adjacent or remote regions of the polypeptide chain.

The angle and intensity of scattered X-rays after passage through a protein lattice has provided the basis of a useful and reliable method for obtaining accurate measurements of the distance separating the repeating units along the polypeptide chain. Briefly, when X-rays pass through an ordered array of atoms some are deflected by passage into regions of high electron density. These scattered X-rays constructively interfere with one another when the angle of the incident beam is at the Bragg angle, θ (half the angle between the incident and diffracted direction) and the distance, d, between the lattice planes can thus be derived from the Bragg equation

$$n\lambda = 2d \sin \theta \tag{2.1}$$

where λ is the X-ray wavelength (usually 154 pm) and n is an integer. X-ray spacings of the protein α-keratin, for example, shows a repeat distance or periodicity of 500 to 550 pm along the axis of the polypeptide chain. By judicious construction of molecular models to the specified dimensions it has been shown that the most likely structure consists of a coiled chain with 3.6 amino acid residues per turn with the side chains extending outwards from the core. This is the classical α-helix and is a particularly stable form of coiling since the orientation of the amino acids is favourable for the formation of hydrogen bonds between each carbonyl oxygen of the α carbon atoms and amide protons three residues removed. If the fibres of α-keratin are stretched, the X-ray periodicity increases to about 660 pm constituting another stable form of the protein designated β-keratin. The repeat distance of the β form of keratin is similar to that found in a number of fibrous proteins, such as fibroin from silk fibres, which have X-ray spacings of about 700 pm consistent with a more extended coiling of the polypeptide chain than α-helix. Proteins typically arranged in β-conformation usually contain high proportions of serine, alanine and particularly glycine and stability is achieved by hydrogen-bonding between different regions of the polypeptide chains when suitably aligned. No membrane proteins for which an amino acid analysis is available contain inordinate amounts of serine, alanine or glycine suggesting that if β-conformation exists it is likely to be restricted to relatively short regions of the polypeptide chains. Finally, some proteins do not show regular X-ray periodicity and the conformation of the peptide chain is less ordered; this arrangement is referred to as random coil.

The secondary and, especially, the tertiary structures of proteins are dominated largely by the interaction of amino acid side chains with surrounding molecules. The folded arrangement of the polypeptide chain results from the spatial disposition of side chains necessary to achieve a configuration of lowest free energy of the whole molecule in its particular environment. These effects have been well documented for a variety of soluble proteins many of which possess amino acid sequences of predominantly hydrophobic residues. The tertiary structure is designed to accommodate these hydrophobic regions relative to the aqueous environment since the chain is folded in such a way as to bring these regions into contact with other sequences of essentially hydrophobic character. The folding is stabilised by cohesive forces between hydrophobic side chains and the exclusion of water molecules results in the formation of a clathrate (cage-like) structure in the surrounding region. If hydrophobic amino acid side chains are forced into an aqueous environment, the motion of surrounding water molecules is restricted and there is a decrease in entropy. Dickerson and Geis[1] have calculated that for every hydrophobic side chain removed from the aqueous region about 17 kJ of free energy of stabilisation is gained by the protein largely at the expense of the accompanying increase in entropy of water molecules. The influence of solvent molecules on the tertiary structure of soluble proteins is reasonably straightforward but in the case of membrane proteins the situation is immensely complicated by the ill-defined environment in which these proteins are located. For example, some membrane proteins exist in a completely hydrophobic environment, some in a hydrophilic environment while most appear to be exposed to some degree of both. Assuming that the same principles governing tertiary structure of soluble proteins also apply to membrane proteins, it is conceivable that in hydrophobic regions of the membrane the nonpolar amino acid side chains are exposed on the outer surface of the protein and the polar residues are sequestered on the inside. This would certainly explain the difficulty of solubilising membrane proteins, since a radical change in tertiary structure would be required for them to reach thermodynamic equilibrium with the altered environment.

Detailed knowledge of the primary structure of membrane proteins is so far lacking, and the only membrane protein for which a complete amino acid sequence has been determined is cytochrome c. As might be expected from its location on the membrane surface, the conformation of cytochrome c is similar to other proteins adapted to polar conditions. There have been attempts to classify membrane proteins subjectively, on the basis of certain physical characteristics, into intrinsic and extrinsic categories with respect to their relative position in the membrane. Capaldi and Vanderkooi[2] for example, have examined the amino acid composition of seventeen membrane proteins and compared these with over 200 soluble proteins. They calculated a percentage polarity score for each protein according to the proportion of amino acids with hydrophilic side chains and found that water-soluble proteins were distributed in a narrow range of 46 ± 6 per cent polarity with a skew towards increasing polarity. Intrinsic membrane proteins, characterised by their resistance to extraction from membranes, poor solubility in water, pronounced surface-active properties and tendency to reform membrane structures on interaction with lipids, invariably scored less than 40 per cent polarity. Extrinsic membrane proteins, on the other hand, resembled soluble proteins and usually scored greater than 45 per cent polarity. All calculations were weighted to include half the total number of neutral amino acid residues, but since these usually account for between 20 and 40 per cent of all residues the calculations will be of only limited value in predicting how membrane proteins are adapted to their particular environment.

2.1.2 Optical Absorption Methods for Assessing Membrane Protein Conformation

All proteins absorb infrared radiation and produce spectra that are characteristic of their particular secondary conformation. Lipids also absorb in the infrared region so that in order to recognise conformational forms of protein in intact membrane preparations a region of the spectrum where lipid absorbance is low must be selected. Two main protein bands, referred to as amide I and amide II, located at frequencies of about 1652 cm^{-1} and 1535 cm^{-1}, respectively, are conventionally used for this purpose. Since sphingolipids also absorb weakly in this region, some caution is warranted when interpreting spectra of membranes containing significant amounts of these lipids. Polypeptides in α-helix or random-coil configurations both produce strong amide I bands but this band is shifted to a frequency of 1630^{-1} for proteins in β-conformation. It is possible to distinguish random coil from α-helix and β-conformation by comparing spectra of membranes prepared in H_2O with those prepared in D_2O, since deuterium substitution causes a selective shift of 10 to 15 cm^{-1} downfield in the position of the amide I band in disordered structures while having no effect on this band from protein in ordered conformation. The amide II band, which is common to all three conformational states, is also eliminated from spectra of random coil polypeptides and, on complete deuterium substitution, a new band appears at about 1450 cm^{-1}.

Protein conformation in erythrocyte membranes and myelin has been investigated by infrared spectroscopy[3,4]. Spectra derived from completely or partially hydrated membranes show strong symmetrical amide I bands about 1652 cm^{-1}, but virtually no trace of a peak at 1630 cm^{-1} to indicate the presence of any β-conformation. An attempt to demonstrate that membrane proteins are not unique in that they can be converted to β-structure by thermal denaturation showed, however, no differences in the characteristic amide I or

II bands in membrane preparations maintained at temperatures ranging from 150° to − 150°C. When the residue remaining after complete extraction of lipids with organic solvents was examined a broad amide I band was observed, centred about 1652 cm^{-1}, with a pronounced shoulder at 1628 cm^{-1} characteristic of β-conformation. Moreover, unless all traces of lipid were removed by the extraction procedure, a structural transition from α-helix to β-conformation could not be demonstrated even by heating to 150°C suggesting that residual lipids that remain tightly bound to the protein exert a stabilising effect on the structure.

The absence of appreciable β-conformation has also been reported in a number of other membrane preparations, with the exception of the inner mitochondrial membrane in which the infrared spectrum indicates a mixture of β-conformation and random coil. The problem of estimating the relative proportions of α-helix and random-coil configurations was examined in erythrocyte membranes by Steim[5] who found a shift in the amide I band from 1651 cm^{-1} to about 1640 cm^{-1} and a corresponding decrease in the amide II band at 1540 cm^{-1} when he transferred membranes from H_2O to D_2O. These changes were consistent with the presence of a considerable amount of disordered structure in erythrocyte membrane proteins with most chains freely accessible to water.

Protein chromophores undergo electronic transitions under ultra-violet radiation and the spectrum of proteins in this region is dominated by a strong absorption band at 190 nm arising from a nonbonding–antibonding transition in the π orbitals about the peptide bond. This absorption band is characteristic of all peptide bonds and it is not particularly sensitive to differences in secondary structure of the protein. Proteins in α-helical conformation, however, have additional absorption shoulders at wavelengths of 205 nm and 220 nm due to a hyperchromic effect on the peptide absorption band, which distinguishes this structure from β-conformation and random coil. The magnitude of the hyperchromic effect, when corrected for absorption by aromatic amino acid side chains, can be used to estimate the percentage α-helix in proteins. Although absorption measurements of membrane proteins in the ultraviolet region have occasionally been used to determine membrane protein conformation[6], the optical activity of these chromophores is much more sensitive to protein secondary structure and is considered the method of choice.

2.1.3 Conformation of Membrane Proteins Determined by Optical Activity

Optically active chromophores rotate the direction of the electric vector when a beam of monochromatic, linearly polarised light passes through a solution of proteins or a suspension of membranes. Unlike absorption spectroscopy, the optical activity originates almost entirely in protein conformation and not in lipid or carbohydrate structure so it is possible to assess protein structure directly in intact membrane preparations[7]. Plane-polarised light can be resolved into two in-plane circularly polarised beams whose vectors rotate in opposite directions. If the frequency of the light is within an optically active absorption region of particular chromophores the light beam will be rotated as well as elliptically polarised (depolarised) after transmission through the sample. Optical activity can be measured either by observing the extent of this rotation to the left or right or by measuring the difference in absorbance of left and right circularly polarised light. When rotation effects are expressed as a function of wavelength a curve of optical rotatory dispersion (O.R.D.) is obtained. Typical O.R.D. curves for poly-L-lysine in α-helix, β-conformation and random-coil arrangements are shown in figure 2.1a. The curves bisect the axis of zero rotation at a wavelength near the ultraviolet absorption maximum and the behaviour of the O.R.D. curve in this region is referred to as the Cotton effect.

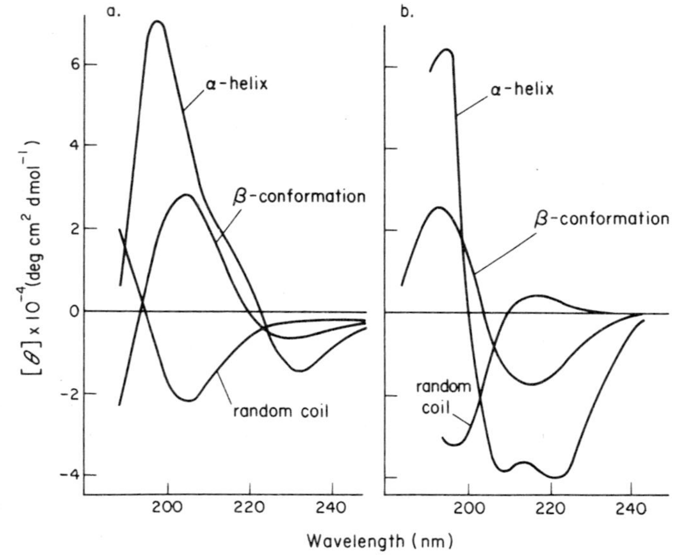

Figure 2.1 Computed curves of optical rotatory dispersion (a) and circular dichroism (b) for synthetic polypeptides in defined secondary conformation.

The Cotton effect is positive if the extremum on the longer-wavelength side of the crossover is positive and the converse when this is negative. The nature of these Cotton effects is used to assess protein conformation since, for example, the depth of the trough of the O.R.D. curve at 233 nm is assumed to be a linear function of helicity of polypeptides. The extent of rotation is usually converted into molar units of rotation, $[\phi]$, from the relationship

$$[\phi] = \frac{\alpha M}{100\, cd} \text{ deg cm}^2 \text{ decimole}^{-1} \qquad (2.2)$$

where α is rotation in degrees to the left or right; M is the relative molecular mass and c the concentration of solute molecules (in g ml^{-1}) and d is the path length of light in decimetres. Optical rotation can be considered as the difference in refractive index for left and right circularly polarised light so the O.R.D. curve is really a measure of the unequal transmission velocity of the light wave. In most samples, appropriate corrections must be applied to account for local field effects arising from the polarisability of solvent molecules, since this can often lead to serious distortions in the O.R.D. spectrum.

Optical activity can also be evaluated by measuring the differences in absorption of left and right circularly polarised light which, when related to wavelength of light, produces a curve of circular dichroism (C.D.). Differences in absorption when expressed as molar absorption coefficients can be converted directly into ellipticity, (θ), which has the same units as optical rotation. Figure 2.1b shows circular dichroism of poly-L-glutamic acid in α-helix and random-coil and poly-L-lysine in β-conformation. Ellipticity is a function of both the magnetic and electric vectors of the incident radiation and the rotational strength of dichroism, (R_k), represents the scalar product of these dipoles. In other words $R_k = \mu_m \cdot \mu_e$ where μ_m and μ_e are the magnetic and electric dipole vectors, respectively.

It will be recalled from the ultraviolet absorption spectrum of protein chromophores that the main electronic transition associated with the peptide bond ($\pi \rightarrow \pi^*$, perpendicular) was located at a wavelength of about 190 nm. There are, however, two other transitions visible in the circular dichroism of proteins in α-helix, one at 208 nm ($\pi \rightarrow \pi^*$, parallel) and another involving the n orbital of the carbonyl oxygen atom ($n \rightarrow \pi^*$), which is observed as a small long-wave tail in the absorption spectrum of α-helix in the region of 220 nm. It should be noted that for these chromophores to be optically active they must be incapable of conversion into their mirror images; otherwise the net dipole effects cancel each other out. Although the peptide bonds in proteins are symmetrical, they are nevertheless perturbed by virtue of their asymmetric disposition in the polypeptide chain and hence are able to satisfy the criteria for optical activity.

Protein structure has been examined by O.R.D. and C.D. methods in plasma membranes from a variety of cell types as well as membranes of the endoplasmic reticulum and mitochondria. As one might expect from a mixture of proteins in different conformations, the spectral position, shape and amplitude are intermediate between those of synthetic polypeptides, as illustrated in figure 2.1. The general conclusion from such studies is that membrane proteins do possess regions of α-helix but absolute amounts of α-helix, β-conformation or random coil cannot be derived with any degree of precision. On the assumption that there is no appreciable β-conformation in membrane proteins, semi-quantitative estimates of α-helical content usually vary between 30 and 50 per cent: values up to 60 per cent and as low as 10 per cent have, however, occasionally been reported. The validity of the underlying assumption used in these calculations cannot be assessed from spectral considerations alone because a considerable amount of β-conformation may be present without being obvious from the spectrum. Consequently other methods, such as infrared absorption spectroscopy, are required to confirm the absence of β-conformation. Studies of inner mitochondrial membranes have indicated that membrane phospholipids are concerned in maintaining protein structure. Masotti *et al.*[8] obtained C.D. spectra from intact and lipid-depleted membranes and found that removal of lipids leads to an extensive alteration in protein conformation. When various purified phospholipids were added to lipid-depleted membranes, the original spectral characteristics were partially restored; the best results were achieved with cardiolipin.

A consistent feature of O.R.D. and C.D. spectra of membranes is their distorted shape when compared with those derived from pure polypeptides in free solution even when solvent effects are taken into account. Typically, these effects are manifested as a decrease in signal amplitude and a slight shift to longer wavelengths at the extremes of the curve. Various interpretations that have been placed on this anomalous behaviour include effects arising from the interaction of neighbouring α-helical regions of the polypeptide, and interaction of the protein with membrane lipids. These conclusions are partly based on the observation that treatment of membranes with phospholipase a or lysophosphatidylcholine removes spectral distortions, though phospholipase c treatment appears to be without effect. Some workers, however, have expressed reservations about accepting such interpretations and they point to the fact that the spectrum of lipid-free membrane proteins still shows anomalous characteristics if the proteins are permitted to aggregate. In a careful examination of the sources of spectral distortions, Urry[9] and his colleagues found that these are largely optical artifacts arising from the particulate, turbid nature of the sample and are superimposed on the optical activity of the protein. They were able to show, for example, that a decrease in amplitude can result from light scatter giving a reduction in the number of chromophores from which a signal is received. These effects are particularly noticeable in membrane suspensions but if precautions are taken to reduce light scatter or, alternatively, to apply suitable corrections for reduction of light transmission, the amplitude of the optically active bands can be restored to normal intensities. Similarly, differential scattering of left and right circularly polarised light was thought to be the main factor responsible for the red shift at the extrema of the O.R.D. and C.D. curves and to be the primary source of C.D. spectral broadening. By applying suitable corrections to C.D. spectra, improved values of ellipticity were obtained at 224 nm and 192 nm for five different membranes (see table 2.1). Both the ellipticity and absorption data indicate the presence of more helical conformation in erythrocyte, mitochondrial and plasma membranes, all of which are associated with a variety of enzymic processes. In contrast, the protein of axonal membranes, where the primary function is ion transport, appears to be arranged in a more extended conformation.

Table 2.1 Circular dichroism and extinction coefficients of some mammalian membrane preparations corrected for spectral distortions

Membrane	Ellipticity $[\theta] \times 10^{-4}$		Molar absorption coefficient ($\times 10^{-3}$)
	224 nm	192 nm	192 nm
human erythrocyte	1.70	3.37	8.7
beef heart mitochondria	1.65	3.22	7.6
rat liver plasma membrane	1.56	2.92	9.0
sarcotubular vesicles	1.08	1.87	7.5
axonal membranes	0.89	1.13	12.6

Data from Urry[9]

2.2 Lipid Structure

Membrane lipids, because of their amphipathic character, interact co-operatively with one another in the presence of water; studies of this associative behaviour have provided a useful basis on which their function in biological membranes can be evaluated. The interaction between lipids can be demonstrated quite simply by shaking a dried membrane lipid extract or purified phospholipid fraction in dilute salt solution. The resulting suspension is turbid and a particulate nature can be surmised from its marked light-scattering properties. When suitably stained preparations are examined under the electron microscope, structures consisting of a large number of concentric layers similar to myelin are observed. If more vigorous methods are employed to disperse phospholipids in water, such as irradiating the suspension with high-energy ultrasonic radiation, the large multilayered structures are broken down into smaller particles consisting of a single layer, which is similar in appearance to the individual layers of the original multilayered structure. The size of these particles cannot be reduced by prolonged ultrasonic treatment and no intact phospholipid molecules are ever found in free solution.

Probably the most important feature of membrane lipids is that they are amphipathic, a property that can be regarded in simple terms as the ability to self-emulsify. This contrasts with the behaviour of nonpolar lipids like triglycerides in aqueous solution, which coalesce to form large aggregates ultimately segregating from the aqueous phase. Completely polar molecules will, of course, dissolve as monomers in water. An intermediate class of compounds is also recognised: these exist as monomers in free solution up to a particular concentration – the critical micelle concentration – beyond which they achieve an equilibrium with a micelle form of the solute. Some membrane components, notably the lysophosphatides, are of this type but they are never found in great abundance in biological membranes and, because of their detergent-like action, they are usually considered to be a source of instability in membranes.

2.2.1 The Lipid Bilayer

The thickness of lipid layers in structures formed by dispersing membrane lipids in aqueous solvents has been measured from electron micrographs of stained preparations and by X-ray diffraction analysis. The latter method is considered the more reliable because measurements can be performed without preliminary treatments that might alter the dimensions of the structure. Shipley[10] has recently reviewed the techniques available for X-ray analyses of lipid structures. As with proteins (see section 2.1.1) X-ray spacings can be derived from the Bragg equation (equation 2.1): however, the distances separating diffraction planes in lipid structures are large relative to the wavelength of the X-rays, and consequently the angle 2θ must be correspondingly small. In other words, the thickness of the lipid layer is considerably greater than 154 pm (the X-ray wavelength commonly employed) and only X-rays scattered from a low-angle incident beam provide an appropriate interference.

Measurements of X-ray long spacings (thickness) are greatly simplified with multilayer structures because the diffracted X-rays from a number of colinearly orientated layers become concentrated into sharp Bragg reflections. For this reason orientated biological membranes like myelin and retinal-rod outer-segment membranes lend themselves well to X-ray diffraction analysis. X-rays scattered by single-layer structures, on the other hand, appear as broad continuous diffraction bands centred about the original diffraction maxima. Nevertheless, some significance is attached to obtaining meaningful X-ray diffraction spacings from single-layered vesicles because most biological membranes are not organised into multi-layered structures. Wilkins *et al.*[11] have derived corrected scattering intensities from a sonicated dispersion (15 per cent water by weight) of egg-yolk phosphatidylcholine and, from the relationship with the angle of scatter, the curve shown in figure 2.2 was obtained. It can be seen that the most intense scattering band is consistent with a long spacing, D, of approximately 3.6 nm with weaker bands corresponding to $D/2$ and $D/3$.

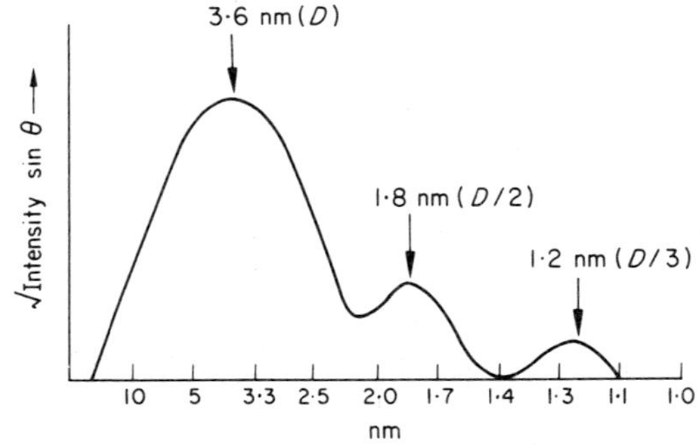

Figure 2.2 X-ray diffraction intensities from single-layered vesicles of phosphatidylcholine (from reference 11).

The relative atomic positions within the lattice planes can be determined from analysis of the intensity distribution between the various diffraction bands by means of a Fourier relationship, which equates electron-density distribution with the diffraction pattern. A Fourier synthesis of X-ray reflections from multilayers of phosphatidylcholine deposited on a mica surface is illustrated in figure 2.3. When attempting to equate electron density with molecular structure not only is the amplitude of the scattered X-rays important (amplitude is proportional to the square root of the intensity) but a knowledge of their phase is also required. There are no precise methods for assigning the correct phase and usually any of a number of phase permutations are possible. Nevertheless, the arrangement of molecules in bimolecular layers as depicted in figure 2.3 is consistent with phase interpretations adopted by these workers and has not been seriously challenged by information derived from other sources.

The overall dimension (repeat distance) of hydrated bilayers (21 per cent water) was about 5.2 nm and the peak-to-peak distance, D, across the centre of the bilayer was 3.68 nm, in close agreement with the spacing of intensity maxima observed in sonicated dispersions of phospholipid shown in figure 2.2.

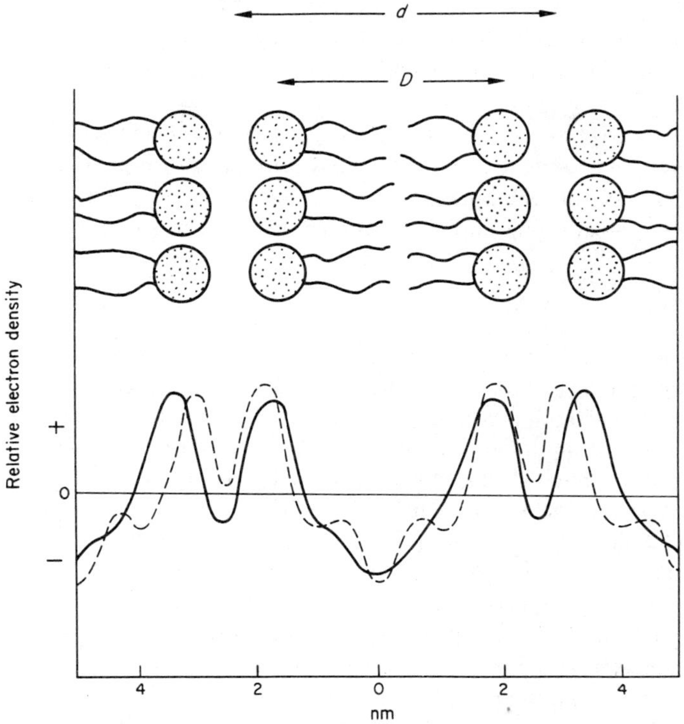

Figure 2.3 Fourier-transformed electron-density profiles in a direction perpendicular to the plane of phosphatidylcholine layers. Broken line: 14 per cent water, 57 per cent relative humidity. Solid line: 21 per cent water, 100 per cent relative humidity. The molecular interpretation corresponding to the electron-density profile (solid line) is shown schematically; circles represent phospholipid headgroups and the lines refer to hydrocarbon chains (data from reference 12).

Regions of electron density greater than water are assumed to be phosphate atoms and regions of lower electron density are associated with methylene ($—CH_2—$) residues of the hydrocarbon side chains. The lowest electron density is located in the centre of the bilayer and is the region where terminal methyl residues of phospholipid hydrocarbon chains from both leaflets of the bilayer reside. When the water content is reduced to 14 per cent by weight, the thickness of the bilayer decreases by 250 pm but the intervening aqueous region separating juxtaposed bilayers is reduced by about twice this amount. It is of considerable interest to see that the hydrocarbon region (that is the distance separating glycerol residues on opposite sides of the bilayer) increases in thickness from 3.68 nm to 3.96 nm and that a pronounced trough appears in the region of the terminal methyl residues. These results strongly suggest that dehydration brings about a more orderly arrangement of the hydrocarbon chains and localisation of the terminal methyl groups at the centre of the bilayer. The effect is consistent with changes in the motion of the acyl chains in the presence of increasing amounts of water as deduced from differential scanning calorimetry (see section 2.3). Sharp reflections are also observed in phosphatidylcholine bilayers when the incident X-ray beam is at a high angle relative to the plane of the bilayer, and these correspond to a lattice spacing of

420 pm. This distance is characteristic of densely packed hydrocarbon chains in close parallel array and indicates that the acyl chains of the phospholipid molecules in the bilayer are fully extended in a direction perpendicular to the bilayer surface.

From a knowledge of the X-ray long spacing across the bilayer in a particular lipid–water system and the density of the lipid component it is possible to calculate the volume of each lipid molecule in the bilayer and hence the area, S, occupied by the polar headgroup in the surface, from the relationship

$$S = \frac{2M}{D\rho N_0} \tag{2.3}$$

where M is the molecular weight of the lipid of density, ρ; D is the X-ray long spacing of the lipid layer and N_0 is Avogadro's number. Measurements derived from the data of Levine and Wilkins[12] for phosphatidylcholine bilayers give an area per molecule for hydrated speciments (21 per cent water) of 0.627 nm^2, decreasing to 0.589 nm^2 when the water content is reduced to 14 per cent. The hydrated packing density is identical with the area occupied by egg-yolk phosphatidylcholine molecules in fully compressed (approximately 43 $mN\ m^{-1}$) monomolecular films orientated at the air–water interface (see section 2.2.2).

The arrangement of hydrated amphipathic lipid molecules in a bilayer structure is particularly favourable from a thermodynamic standpoint, since water is excluded from the interior of the bilayer where the hydrocarbon residues are located and the polar headgroup is afforded maximum exposure to the aqueous phase on the outside. The foundations for assuming the existence of such arrangement of lipids in biological membranes were laid more than 100 years ago by Mettenheimer[13] who first described the double refraction of myelin, and later by Wynn[14] in his studies of the birefringent properties of nerve tissue. Subsequent studies using polarisation microscopy led Schmidt[15] to formulate a structure for myelin that incorporated an arrangement of lipid molecules with their long axes perpendicular to the plane of the membrane. Since then, the lipid bilayer has proved to be a durable arrangement not only in respect of its physical resilience but also in the dogma of membrane structure.

2.2.2 Physical Properties of the Bilayer Hydrocarbon Region

The most remarkable feature of the hydrocarbon residues of membrane lipids is their almost unlimited potential for chemical variability. Each class of phospholipid, for example, consists of a large number of different molecular species, which can be distinguished by the length of the constituent acyl side chains and the number and position of unsaturated bonds. The hydrophobic region of bilayers formed from any particular class of phospholipid will therefore possess characteristics that reflect the particular type and proportion of each hydrocarbon chain represented in the mixture. Experimental approaches to the study of physical properties of the hydrocarbon region have concentrated on examining the behaviour of purified lipids with chemically defined hydrocarbon constituents. These studies have been greatly assisted by recent advances in lipid chemistry enabling synthesis of a range of chemically defined phospholipids in modest yield, so that the laborious procedures required to purify them from biological material can be avoided.

Certain changes in the physical properties of the membrane lipid hydrocarbons on addition of water or as a result of a change in temperature have established the fact that these molecules are mesomorphic and can exist in one of several distinct physical forms. The dependence of these forms on the chemical composition of the lipid has been examined by Luzzati *et al.*[16,17]. Two forms of the lamellar structure are most relevant to membrane structure, an Lα, high-temperature liquid form (figure 2.4b) and an Lβ form in which the hydrocarbon chains are stiff, relatively immobile and packed in hexagonal array (figure 2.4a). These two phases are usually referred to as liquid-crystal

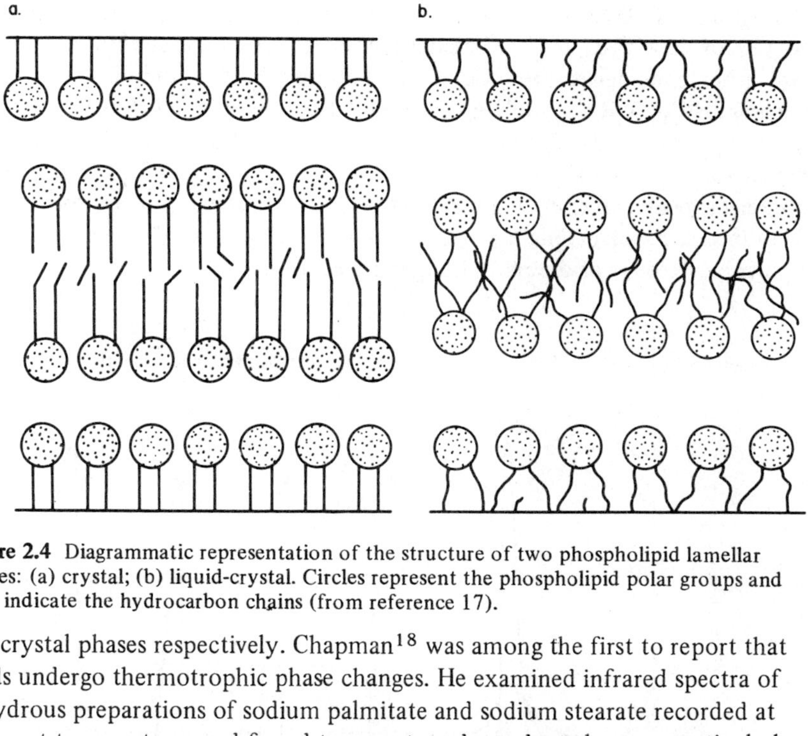

Figure 2.4 Diagrammatic representation of the structure of two phospholipid lamellar phases: (a) crystal; (b) liquid-crystal. Circles represent the phospholipid polar groups and lines indicate the hydrocarbon chains (from reference 17).

and crystal phases respectively. Chapman[18] was among the first to report that lipids undergo thermotrophic phase changes. He examined infrared spectra of anhydrous preparations of sodium palmitate and sodium stearate recorded at different temperatures and found temperature-dependent changes, particularly in the spectral region of 720 cm^{-1}. This absorption band in paraffins is attributed to the interaction between methylene groups of neighbouring hydrocarbon chains and becomes less well resolved as the temperature increases indicating an increase in chain mobility. Transitions from an ordered structure (crystal phase) to disordered structure (liquid-crystal phase) at characteristic temperatures (T_c) have now been reported for a number of phospholipids. In general, for phospholipid molecules of the same class and equivalent hydration states, the longer, more saturated chains have higher transition temperatures than short, unsaturated chains and *trans*-unsaturated bonds undergo phase transitions at higher temperatures than *cis*-unsaturated bonds in equivalent positions.

In subsequent studies by Chapman *et al.*[19–21] the thermotrophic mesomorphism of a homologous series of synthetic 1,2-diacyl-L-phosphatidylcholines with hydrocarbon chain lengths from ten to twenty-two carbon atoms have been examined by infrared spectroscopy, X-ray diffraction and differential scanning calorimetry. The phase transition from crystal to liquid-crystal form is an endothermic reaction and can be readily detected by differential scanning calorimetry. This technique involves measuring the differential heat energy required to maintain a constant rate of temperature change in the sample and then, if the recorded peaks from such experiments are mathematically integrated, it is possible to convert the results directly into calorimetric units. The differential scanning calorimeter curves of 1,2-dimyristoyl (C_{14}), 1,2-dipalmitoyl (C_{16}), 1,2-distearoyl (C_{18}) and 1,2-dibehenoyl-L-phosphatidylcholines (C_{22}) are shown in figure 2.5. The ice-melting profile of each curve can be observed (0°C) and the endothermic transition temperature (T_c), measured at the point of departure of the curve from the baseline, can be seen to increase with increasing hydrocarbon chain length from C_{14} to C_{22}. Phospholipids with hydrocarbon chain lengths C_{12} and C_{10} undergo complex transitions at and below 0°C, respectively. The corresponding heats of transition also increase with increasing chain length from 41 J g^{-1} (C_{14}), 50 J g^{-1} (C_{16}) to 57 J g^{-1} (C_{18}). The small endothermic peak preceding the main hydrocarbon transition and approaching T_c with increasing chain length is believed to coincide with a rearrangement of the polar headgroups of the phospholipid molecules.

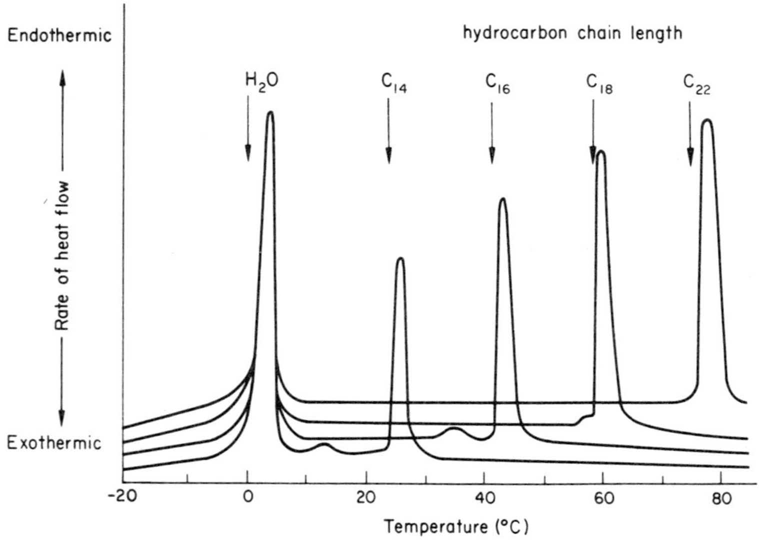

Figure 2.5 Differential scanning calorimeter heating curves for saturated 1,2-diacylphosphatidylcholines dispersed in an equal mass of water (data from references 20, 21).

Phase changes can also be observed from a change in the X-ray short spacing of phospholipid lamellae. The diffraction band at 420 pm associated with hydrocarbon chains packed in an ordered hexagonal lattice is replaced by a diffuse diffraction band corresponding to a spacing of 460 pm when lipids are heated above their phase transition temperature. This suggests a loss of crystallinity at temperatures above the phase-transition temperature and, as a result of the increase in chain mobility, the overall thickness of the phospholipid bilayer is reduced by about 500 pm. Measurements of bilayer volume from X-ray spacings have been confirmed more directly by Traüble and Haynes[22] who recorded volume changes in dipalmitoylphosphatidylcholine bilayers by dilatometry (see figure 2.6). They found a sharp increase of 1.4 per cent in bilayer

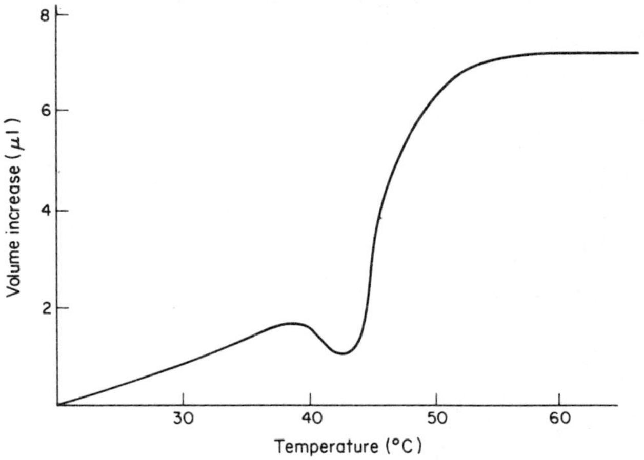

Figure 2.6 Change in volume of a dispersion of 0.5 g dipalmitoylphosphatidylcholine in 2 ml water with temperature (see reference 22).

volume and an increase from 0.48 nm^2 to 0.58 nm^2 in the area occupied by each molecule in the surface of the bilayer on heating from temperatures below to that above the phase-transition temperature. These changes were centred about a temperature of 44°C and reflect a transition from crystalline to liquid-crystalline structure in the hydrophobic region. Since the observed change in volume is about one order of magnitude less than would be expected for melting dilation of completely disordered hydrocarbon chains, they suggested that the liquid-crystalline structure of the bilayer still retains a relatively ordered conformation and never assumes a completely liquid state. Such restrictions in chain motion may permit the creation of small pockets of free volume between the hydrocarbon chains as kinks develop and their effective length is reduced.

Another method whereby molecular volume and acyl-chain mobility can be assessed is to observe the area occupied by each molecule in monomolecular films at the air–water interface as a function of surface pressure. When a solution of phospholipids in a volatile solvent is applied to a clean aqueous surface, the solvent evaporates leaving the phospholipid molecules in the surface layer. Provided that an appropriate number of molecules are added per unit area of surface, the film will be one molecule thick and, because the phospholipid is amphipathic, each molecule will be orientated with the hydrocarbon residues projecting into the air phase and the polar headgroup immersed in water. If a low-density film is compressed by reducing the available surface area, the surface tension decreases – the magnitude of this decrease is referred to as the surface pressure of the monolayer – up to a point where no further decrease in surface tension results from compression of the film. At this point the film collapses and is no longer a monomolecular layer. Phillips and Chapman[23] have recorded surface pressure–area relationships for a series of saturated 1,2-diacylphosphatidylcholines and phosphatidylethanolamines; the curves obtained for the choline homologues are presented in figure 2.7. Dibehenoyl (C_{22}) and distearoyl (C_{18}) phosphatidylcholines are typically condensed films with a limiting area of 0.44 nm^2, whereas the shorter-chain dimyristoyl (C_{14}) and dicaproyl (C_{10}) phosphatidylcholines are considerably more expanded.

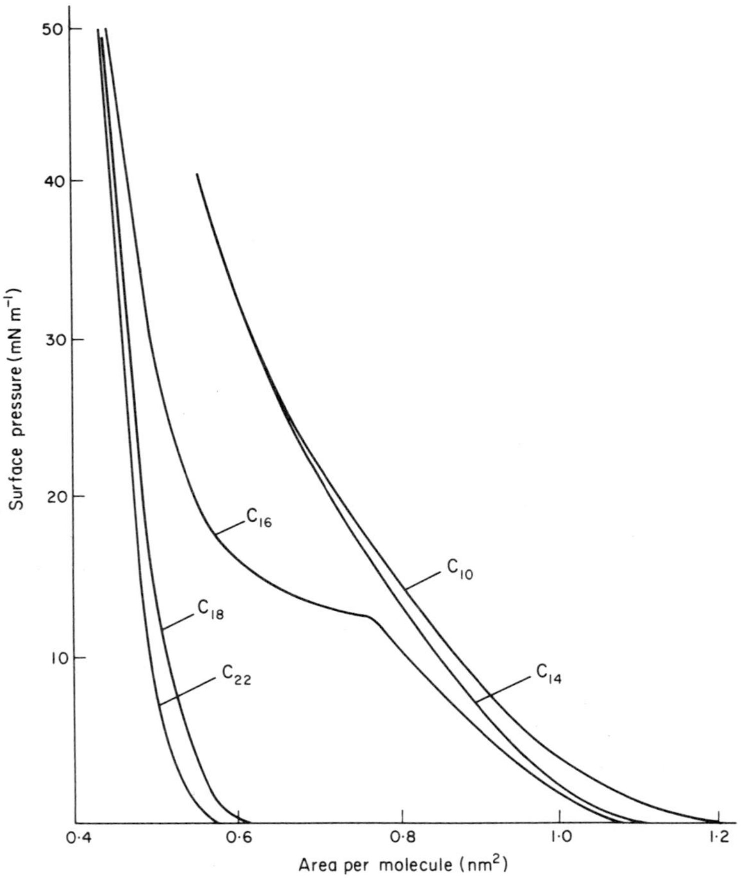

Figure 2.7 Surface pressure – area isotherms for a homologous series of saturated 1,2-diacylphosphatidylcholines spread on a subphase of 0.1M NaCl at 22° (data from reference 23).

Measurements of enthalpies and entropies associated with a transition from condensed to expanded monolayers were consistent with the corresponding thermodynamic parameters associated with crystal to liquid-crystal phase transitions in hydrated bimolecular lamellae, suggesting that the respective condensed and expanded states were analogous in the two structures. Dipalmitoylphosphatidylcholine exists in two forms at 22°C depending on the surface pressure of the monolayer; at pressures less than 10 mN m^{-1} it is an expanded type of monolayer, but above this pressure there is a transition to a more condensed form. The pressure at which this transition is observed is related to the temperature so that at 6°C the film is condensed at all pressures and at 35°C it is expanded at least up to a pressure of 30 mN m^{-1}. It was suggested that at a temperature of 41°C the monolayer would be expanded even at the collapse pressure. These observations together with surface-area calculations from X-ray and other studies indicated that phospholipid molecules are very closely packed in the bimolecular leaflet, probably occupying areas equivalent to the limiting area observed in monomolecular films.

The nature of the polar group of particular phospholipid classes appears to have some influence on the hydrocarbon phase transition. Surface pressure–area curves of saturated diacylphosphatidylethanolamines are more condensed than the corresponding choline homologues with a limiting area of only 0.4 nm^2 and the transition from expanded to condensed films at 22°C was observed with the dimyristoyl (C_{14}) phospholipid rather than the C_{16} phospholipid found in the choline series. Taylor *et al.*[24] recorded the surface pressure–area curve of a mixed film of dipalmitoylphosphatidylcholine and dipalmitoylphosphatidylethanolamine spread at a water–iso-octane interface, where the interaction between hydrocarbon chains of adjacent film molecules is considerably diminished while having little effect on the interaction between polar headgroups. The behaviour of the mixed film was the same as would be expected from the additive effects of the two pure phospholipids, suggesting that the molecules segregate in the film according to their particular phase-transition temperature rather than becoming intimately mixed.

2.2.3 The Use of Molecular Probes to Study Lipid Structure

Fluorescent, optically absorbing and spin-label probe molecules have all been used to study phospholipid mobility in bilayer dispersions. Sackmann and Traüble[25] for example, have used the fluorescent probe, 8-anilino-l-naphthalene sulphonate (ANS) and the pH indicator, bromothymol blue, to detect changes in the polar headgroup arrangement in sonicated bilayer vesicles of dipalmitoylphosphatidylcholine undergoing thermotrophic phase transitions. ANS is an amphipathic molecule consisting of a hydrophobic ring structure attached to a charged sulphonic acid group and is likely to orientate in the bilayer with the charged group residing in the region of the polar headgroups of the phospholipids and the fluorescent residue extending into the hydrophobic interior. The quantum yield, Φ, of the ANS probe is a function to the polarity of its environment and increases from a value of $\Phi = 0.004$ in water to $\Phi = 1$ in apolar solvents. When ANS is added tc phospholipid dispersions the quantum yield increases to $\Phi = 0.08$ indicating that the probe extends at least partly into the hydrocarbon region of the bilayer: binding is also accompanied by a shift in the wavelength of maximum emission from 520 nm in water to 485 nm in the bilayer. Bromothymol blue also binds to phospholipid bilayers by virtue of its aromatic ring structure, and binding causes a decrease of nearly two orders of magnitude in the molar absorption coefficient ($\epsilon_{615\ nm} = 1.4 \times 10^4$ in water at pH 7; $\epsilon_{615\ nm} = 2.4 \times 10^2$ in the bilayer). The fluorescence at λ_{485} and absorbance at λ_{615} of probe molecules added to bilayer dispersions of dipalmitoylphosphatidylcholine are shown in figure 2.8 and marked changes about the phase-transition temperature of the phospholipid can be seen. Sackmann and Traüble[25] titrated probe molecules against the phospholipid dispersion at temperatures above and below the phase-transition temperature and concluded that there was a threefold increase in the number of probe-binding sites on transition from crystal to liquid-crystal phase. This suggests a loosening of the polar-headgroup structure with a lateral expansion of the bilayer when the hydrophobic region assumes a liquid-crystal conformation.

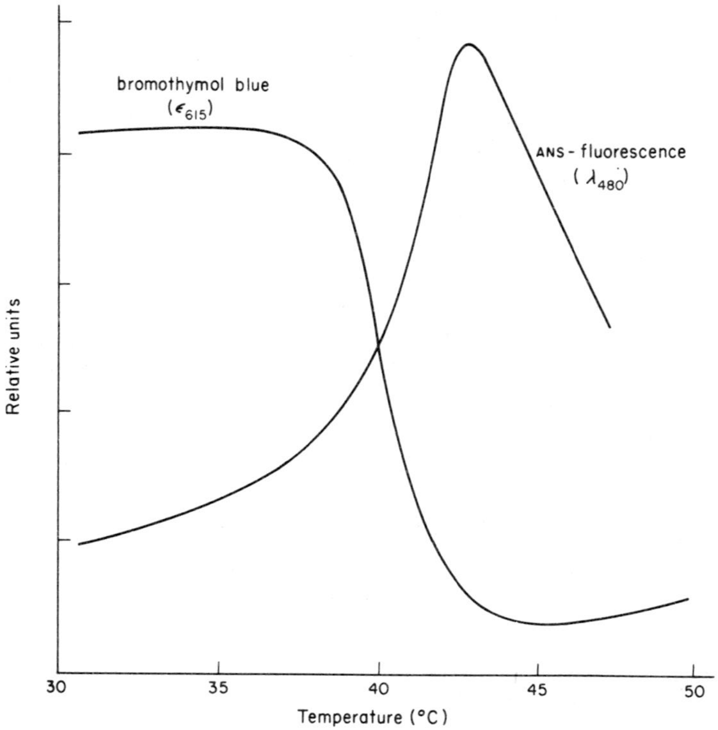

Figure 2.8 Changes in fluorescence of 8-anilino-l-naphthalene sulphonate (ANS) and absorbance of bromothymol blue added to an aqueous dispersion of dipalmitoylphosphatidylcholine at different temperatures (after Sackmann and Träuble[25]).

Fluorescent-probe techniques have also been employed to measure the motion of hydrocarbon chains in the bilayer interior. The probes used in such experiments consist of predominantly hydrophobic residues, which enables them to penetrate completely into the centre of the bilayer. Their structure (two examples are shown in figure 2.9) precludes any preferential binding to the polar headgroups or the hydrocarbon chains of phospholipid molecules and the available evidence suggests that they are evenly distributed throughout the hydrophobic region of the bilayer. The motion of fluorescent probes can be examined by fluorescence depolarisation techniques in which the local viscosities opposing in-plane and out-of-plane rotations of the probe molecule can be estimated. A description of the instrumentation required to obtain depolarisation measurements and the theory involved can be found in a recent paper by

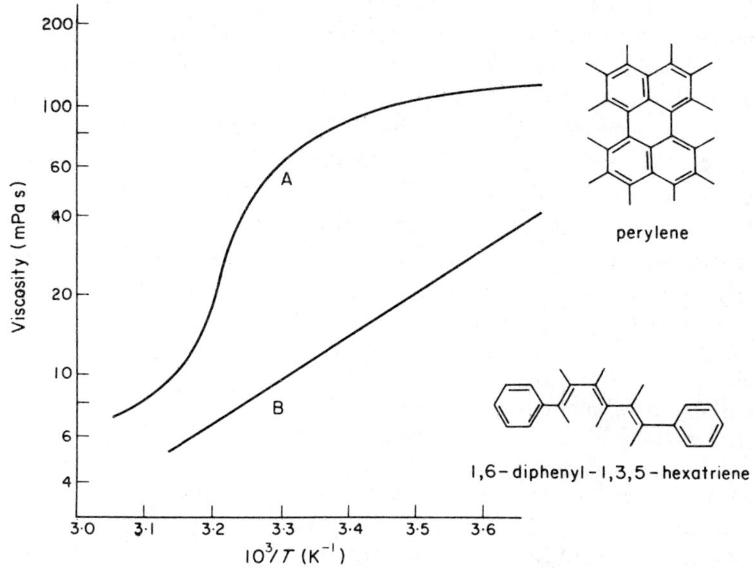

Figure 2.9 Microviscosity of dipalmitoylphosphatidylcholine (A) and yeast phosphatidylcholine (B) as a function of temperature. Curve A was obtained with a perylene probe and curve B with 1,6-diphenyl-1,3,5-hexatriene (from references 27, 28).

Shinitzky *et al.*[26]. In order to obtain absolute viscosities, each fluorescence probe must be calibrated by measuring the fluorescence anisotropy of the dye in an immobilised state (r_0) relative to that in a mobile state (r), usually in oil at various temperatures. Viscosity is derived from the relationship

$$\frac{r_0}{r} = 1 + \frac{kT\tau}{\eta V_r} \tag{2.4}$$

where k is the Boltzmann constant, T is the absolute temperature, τ is the lifetime of the probe in a fluorescent state (see Cogan *et al.*[27] for procedures used to measure fluorescence decay), η is the viscosity of the probe environment and V_r is the effective molecular volume of the rotating dye. This relationship approaches linearity with spherical probe molecules but can deviate substantially for non-spherical probes. Rotation of the probe molecule in the hydrocarbon region of lipid bilayers measured as fluorescence anisotropy, r, is obtained from light intensities, I, received through a polariser oriented parallel ($I_{\parallel}$) and perpendicular ($I_{\perp}$) to the plane of polarisation of the exciting beam, where

$$r = \frac{I_{\parallel} - I_{\perp}}{I_{\parallel} + 2I_{\perp}} \tag{2.5}$$

after applying the necessary corrections to account for light scatter in the sample. The value of r at any temperature can be used to obtain viscosity from the calibration curve of the particular probe.

The extent of fluorescence depolarisation of perylene in dipalmitoylphosphatidylcholine dispersions between 0°C and 50°C has been determined by Cogan *et al.*[27] (see curve A of figure 2.9). There is a pronounced decrease in viscosity with increasing temperature in the range 25°C to 50°C with a midpoint close to 39°C. Viscosities of bilayer preparations of egg-yolk phosphatidylcholine or yeast phosphatidylinositol (curve B of figure 2.9), both possessing highly unsaturated hydrocarbon chains, do not deviate from the exponential form $\eta = A e^{\Delta E/RT}$, characteristic of pure liquids. A is the fraction of total emission intensity given by the fluorescent component. The fusion activation energy, ΔE, can be calculated from the slope of the log η versus $1/T$ plot ($\Delta E/R$) where R is the gas constant and T is the absolute temperature. Linearity of the curve indicates that bilayers of the two unsaturated phospholipids are in a liquid-crystal phase in the temperature range 0–50°C and ΔE is ~ 38 kJ mole^{-1}.

Another probe technique that has been used sucessfully to obtain information about local polarity and viscosity in lipid bilayers is electron spin resonance spectroscopy. The basis of this method is that if an odd-electron molecule is placed in an essentially diamagnetic biological structure, its motion can be determined and, as we have seen in the case of fluorescence probes, this motion can be used to measure physical parameters of the environment surrounding the probe. For an informative account of the origin of magnetic resonance the reader is referred to a monograph by McLauchlan[29]. Smith[30] has reviewed the application of spin–label methods in the study of lipid and membrane structure. Basically, electron spin resonance involves the tendency of a magnetic dipole arising from the spin of an unpaired electron to align in an externally applied magnetic field in a direction parallel (low energy) or antiparallel (higher energy) to the field. Electronic transitions between these two spin states are induced when a second electromagnetic field, coupled to the main field, oscillates at a frequency corresponding to the energy difference between the two spin states. Under these conditions a state of spin resonance is achieved, and because the population in the more stable parallel orientation exceeds those in the antiparallel direction there is a net absorption of energy from the oscillating field. This absorption is the source of the electron spin resonance spectrum.

The most common paramagnetic probe used in biological studies is the relatively stable nitroxide radical, which possesses an unpaired electron in a 2pπ nitrogen nonbonding orbital. The structure of the nitroxide radical is

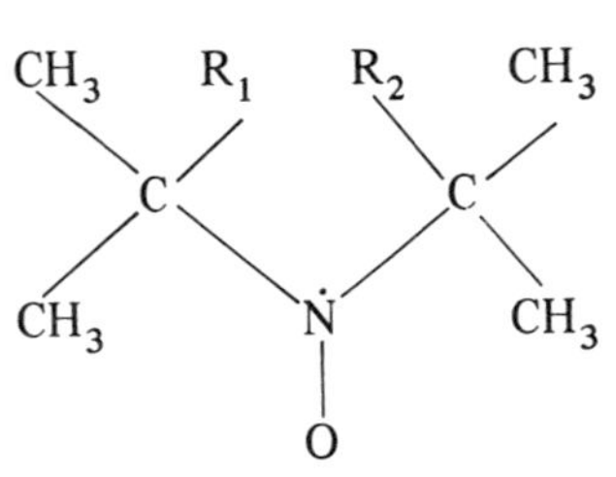

where R_1 and R_2 represent non-specified residues that can be varied to enable the nitroso group to penetrate into the hydrocarbon region of the bilayer or partition into a more hydrophilic environment. Spin–label analogues of phospholipids, cholesterol and other membrane constituents have been prepared and incorporated into lipid bilayers or biological membranes to provide information about the molecular organisation.

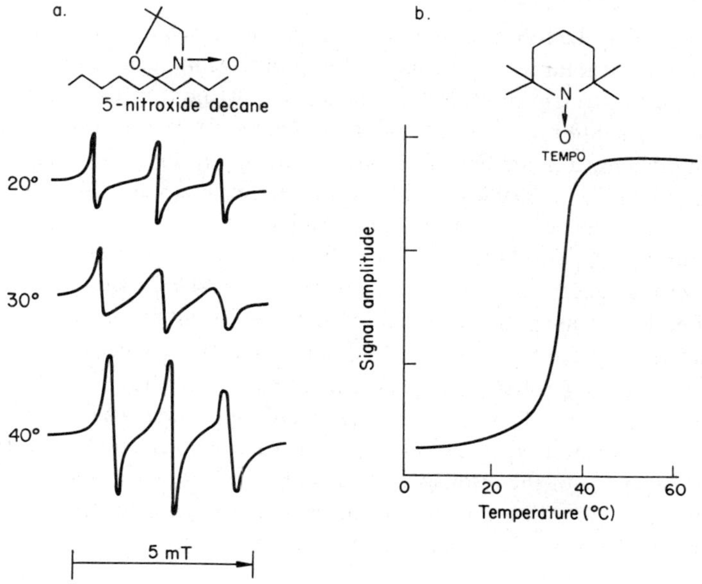

Figure 2.10
(a) Electron spin resonance spectra of 5-nitroxide decane probes in dispersions of dipalmitoylphosphatidylcholine.
(b) Signal amplitude of TEMPO in dispersions of dipalmitoylphosphatidylcholine at different temperatures (from references 31 and 32).

Paramagnetic resonance intensity of the spin probes 2,2,6,6-tetramethylpiperidine-l-oxyl (TEMPO) and 5-nitroxidedecane have been used to detect thermotrophic phase changes in dispersions of dipalmitoylphosphatidylcholine. Typical spectral changes for 5-nitroxidedecane in aqueous dispersions of the phospholipid at three temperatures are shown in figure 2.10a. As the temperature increases so does signal amplitude, and when signal height of the TEMPO probe is plotted as a function of temperature this increases markedly about the phase-transition temperature of the phospholipid (see figure 2.10b). The reason for this is that resonance intensity is relatively low in an aqueous medium but is enhanced when the probe molecules enter the hydrophobic interior of the bilayer at higher temperatures. The spectral characteristics indicate that the probes are tumbling rapidly and that the lipid is likely to be in a liquid-crystal phase.

An alternative procedure has been developed that does not rely simply on a partition of the probe between the aqueous phase and the bilayer. The method involves synthesis of spin-label analogues of membrane lipids, which are then incorporated into bilayers. A number of these analogues have been synthesised, including a series of phosphatidylcholines with paramagnetic groups attached to specified positions of the hydrocarbon chain[32]. The molecular motion of these spin probes in bilayers of dipalmitoylphosphatidylcholine have indicated that the motion of phospholipid molecules is relatively constrained in the region of the glycerol backbone but that the hydrocarbon chains become increasingly flexible and mobile as they extend into the centre of the bilayer.

2.2.4 Nuclear Magnetic Resonance Techniques in Studies of Lipid Structure

One of the principal advantages of nuclear magnetic resonance spectroscopy in studies of lipid structure is that, unlike other probe techniques, the motion of nuclei of the constituent molecules can be assessed without introducing foreign probe molecules into the system, which may themselves perturb the structure. For a comprehensive treatment of the theory and application of nuclear magnetic resonance the reader should consult references 33 and 34. Horwitz[35] has also provided an account of the uses of nuclear magnetic resonance spectroscopy in studies of phospholipid and membrane structure. In principle, nuclei with nonzero magnetic moments will tend to align parallel or anti-parallel to the direction of an applied external magnetic field constituting high and low energy states, respectively. Under the influence of a second electromagnetic field oscillating at a frequency characteristic of particular nuclei, these nuclei will absorb energy and flip over to a higher-energy spin state. The absorption of energy from the radiofrequency field, as in the case of electron spin resonance, serves as the source of the resonance signal and, in the constant-field technique, the sample is swept with variable frequencies, ν, appropriate for the resonance conditions of the particular muclei defined by

$$\nu = \frac{\gamma H_0}{2\pi} \qquad (2.6)$$

where γ is the characteristic gyromagnetic ratio of the particular isotope and H_0 is the magnetic strength of the main field. Isotopes that fulfil the condition of nonzero magnetic moments and have application in membrane studies include ^{1}H, ^{13}C, ^{19}F and ^{31}P. From equation 2.6 it can be seen that frequency is a function of field strength, and consequently improved resolution of individual resonances can be achieved by increasing the external magnetic field strength; 220 MHz is the practical limit in contemporary spectrometers, although instruments operating at field strengths in the 60 MHz and 100 MHz ranges are capable of yielding valuable information. Apart from the external field strength, the local magnetic field experienced by any particular nucleus is modified by dipole interactions with neighbouring nuclei, by its particular covalent bond coupling with nonequivalent nuclei and by shielding effects giving rise to chemical shifts. In proton magnetic resonance, for instance, particular protons in a molecule can be recognised by the magnitude of this chemical shift, which is conventionally referred to the resonance line of externally added tetramethylsilane or an equivalent reference standard.

Chemical shifts of protons observed in the magnetic resonance spectrum of egg-yolk phosphatidylcholine dissolved in deuterated chloroform are illustrated in figure 2.11a. The identities of protons in particular couplings are recognised from their chemical shift with respect to the reference compound (tetramethylsilane) and the number of protons resonating at each frequency is proportional to the area under the absorption band. Sharp absorption peaks are observed when the phospholipid is dissolved in $CDCl_3$, because the resonances of the various protons are relatively unhindered by interaction with neighbouring nonpolar solute molecules. When the same phospholipid is gently dispersed in D_2O, the proton magnetic resonances are considerably broadened and some

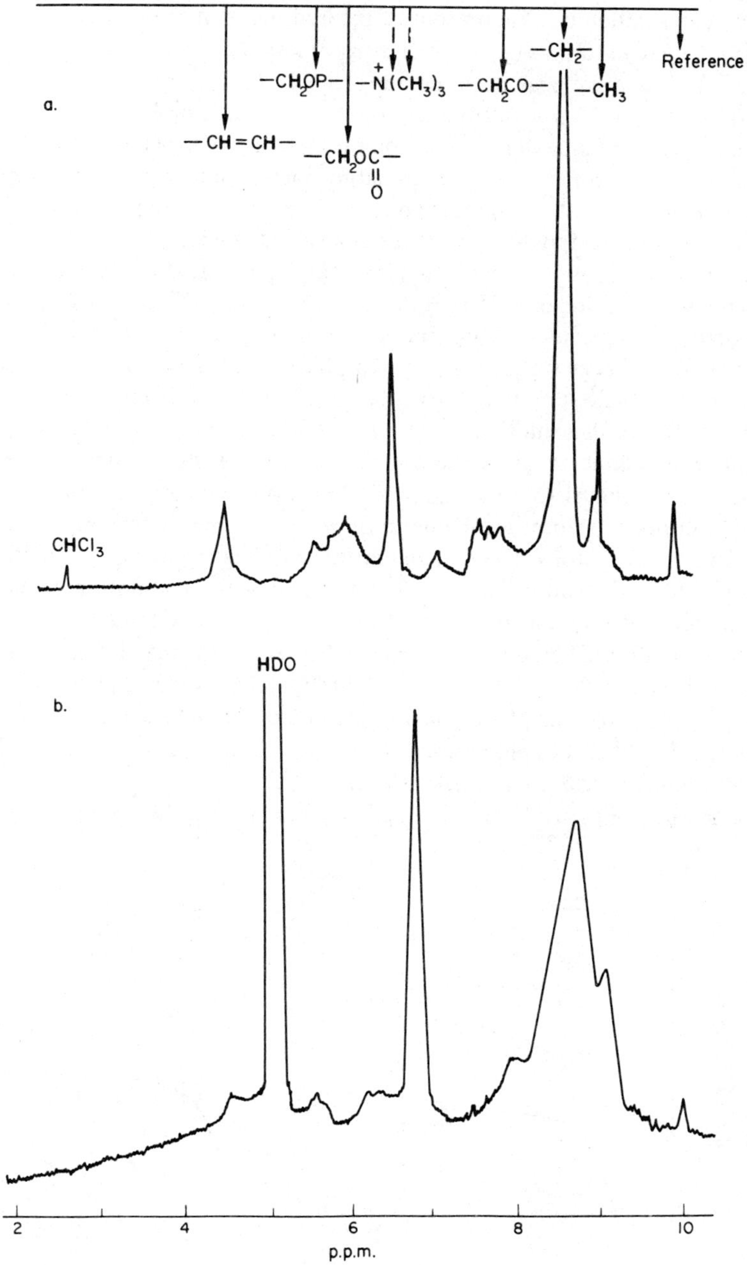

Figure 2.11 Proton magnetic resonance spectra (60 MHz) of egg-yolk phosphatidylcholine: (a) 7 per cent by weight in $CDCl_3$; (b) 5 per cent by weight ultra-sonicated in D_2O. Chemically shifted proton resonances are identified at the top (adapted from references 36 and 37).

bands, including hydrocarbon-chain methylene protons, are not resolved at all. Some improvement in resolution is achieved if the phospholipid is dispersed by ultrasonic irradiation but the protons of the hydrocarbon chains are still broadened relative to those of the quaternary methyl residues of choline (figure 2.11b) and the peak corresponding to $-N^+(CH_3)_3$ protons is shifted upfield by about 0.17 p.p.m. from its original position in molecules in free solution. The shape and width of proton magnetic resonance absorption bands depend on the environment of the resonating nucleus and are influenced by dipole interactions with surrounding nuclei, by magnetic field inhomogeneities within the sample and by lifetime of the spin states. Sheard[38] examined the effects of ultrasonication of egg-yolk phosphatidylcholine and concluded that differential broadening of $-N^+(CH_3)_3$, $-CH_2-$ and $-CH_3$ signals is due to magnetic-field inhomogeneity near and within large particles. Nevertheless, there is some evidence suggesting that phospholipids in small bilayer vesicles are in a more mobile state than those in coarse bilayer dispersions (see Horwitz *et al.*[39]). Hsu and Chan[40] have investigated proton magnetic resonance intensities of choline and hydrocarbon-chain methyl residues in unsonicated preparations of dimyristoyl and dipalmitoylphosphatidylcholines using a delayed Fourier-transform technique to filter out the rapidly relaxing methylene-chain protons. The signal intensities of these protons as a function of temperature is shown in figure 2.12 and indicates that the motion of these protons increases simultaneously about the respective phase transition temperatures although never more than 50 per cent of the maximum intensity is observed. The motion of the choline and terminal methyl protons is therefore restricted even above the phase-transition temperature, in agreement with the suggestion that phospholipid molecules are more densely packed in unsonicated bilayers compared with small single bilayer vesicles.

Lee and his colleagues[41] have used the spectral band width parameter of

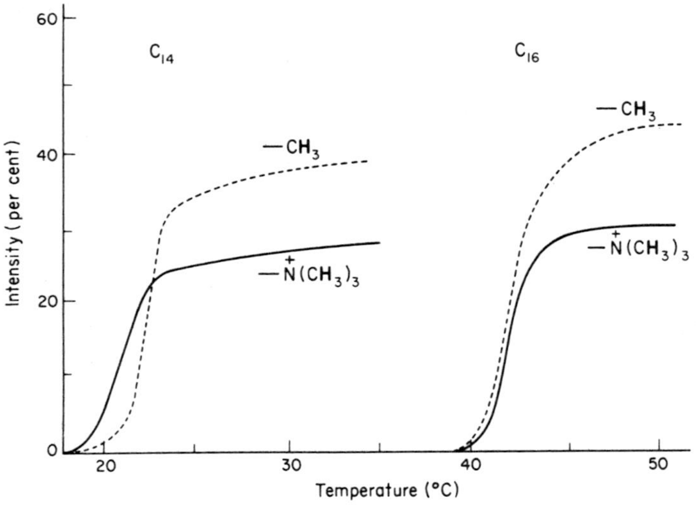

Figure 2.12 Variation of proton magnetic resonance intensity of $-CH_3$ and $-N^+(CH_3)_3$ protons in dispersions of dimyristoyl (C_{14}) and dipalmitoyl (C_{16}) phosphatidylcholines with temperature (from reference 40).

—$N^+(CH_3)_3$, —CH_2— and —CH_3 protons of sonicated dispersions of dipalmitoylphosphatidylcholine to examine the lifetime of spin states above and below the phase transition temperature. Two factors influence the lifetime of a resonating nucleus as it decays to its unexcited energy level; spin–lattice (half-life T_1) and spin–spin (half-life T_2) relaxation modes. Unlike T_1, T_2 is independent of temperature so that variation in proton band width at different temperatures only reflects changes in T_1. Spin–lattice relaxations of —$N^+(CH_3)_3$, —CH_2— and —CH_3 protons of dipalmitoylphosphatidylcholine are not equivalent with respect to their response to increasing temperatures above the phase-transition temperature of the phospholipid.

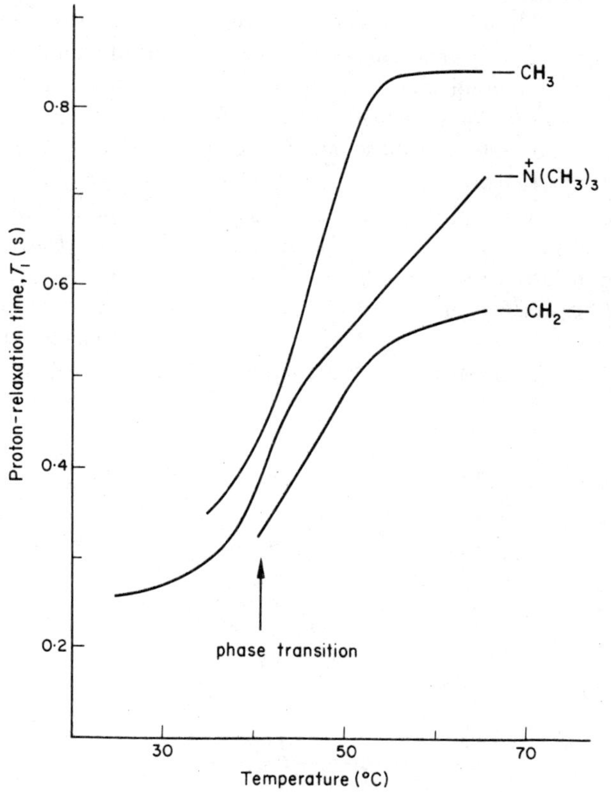

Figure 2.13 Spin–lattice relaxation times (T_1) of different protons in a sonicated dispersion of dipalmitoylphosphatidylcholine as a function of temperature. The slope of these curves reflects differences in the environment of particular protons in the molecule. The phase-transition temperature of the phospholipid is 41°C.

Thus the rate of signal-broadening, represented in figure 2.13 as spin–lattice relaxation time, is different for each particular proton resonance as the temperature is reduced from about 65°C to 40°C, although the area of each band is constant, indicating that the same number of protons is contributing to each signal.This is also true of the —$N^+(CH_3)_3$ resonance signal below the phase-transition temperature, but a marked broadening of this band at about 40°C indicates a sharp increase in T_1 at this point. In contrast, —CH_2— and

—CH_3 proton magnetic resonance signals decrease in area below the phase-transition temperature, eventually disappearing altogether from the spectrum at about 35°C. This effect has been attributed to immobilisation of the acyl hydrocarbon chains as they assume a crystal structure in the interior of the bilayer.

In addition to continuous-wave nuclear magnetic resonance spectroscopy, relaxation times can be obtained by a method known as pulsed nuclear magnetic resonance. The main advantage of pulsed methods is that, unlike continuous-wave spectroscopy, where only average values of spin–lattice relaxation time for equivalent nuclei are obtained, relaxation times for all nuclei producing a resonance signal can be measured independently. The technique involves the irradiation of the sample at fixed field with brief, intense pulses of radio frequency energy in a defined sequence, inducing all specified nuclei to resonate at their characteristic frequency. The resulting free-induction decay of the signal at the end of the pulse sequence can be transformed mathematically to produce spectra equivalent to those obtained by continuous-wave procedures. The method is particularly suited to measuring relaxation times of nuclei such as ^{13}C, which is low in natural abundance (1.1 per cent) and relatively magnetic-resonance insensitive (< 2 per cent of proton sensitivity at equivalent field strength). Spin–lattice relaxation times of ^{13}C nuclei of dipalmitoylphosphatidylcholine dispersed as bilayer vesicles above the phase-transition temperature (figure 2.14) indicate that the glycerol backbone of the molecule is relatively immobilised within the structure (short T_1) with increasing mobility (longer

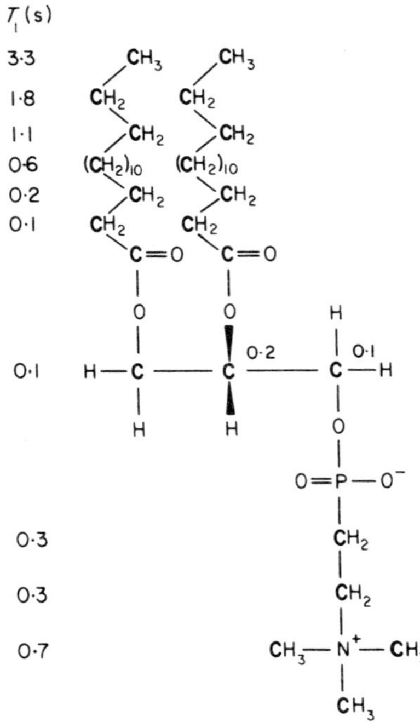

Figure 2.14 Spin–lattice relaxation times (T_1) of ^{13}C nuclei of dipalmitoylphosphatidylcholine dispersed in D_2O at 52°C (data from reference 42).

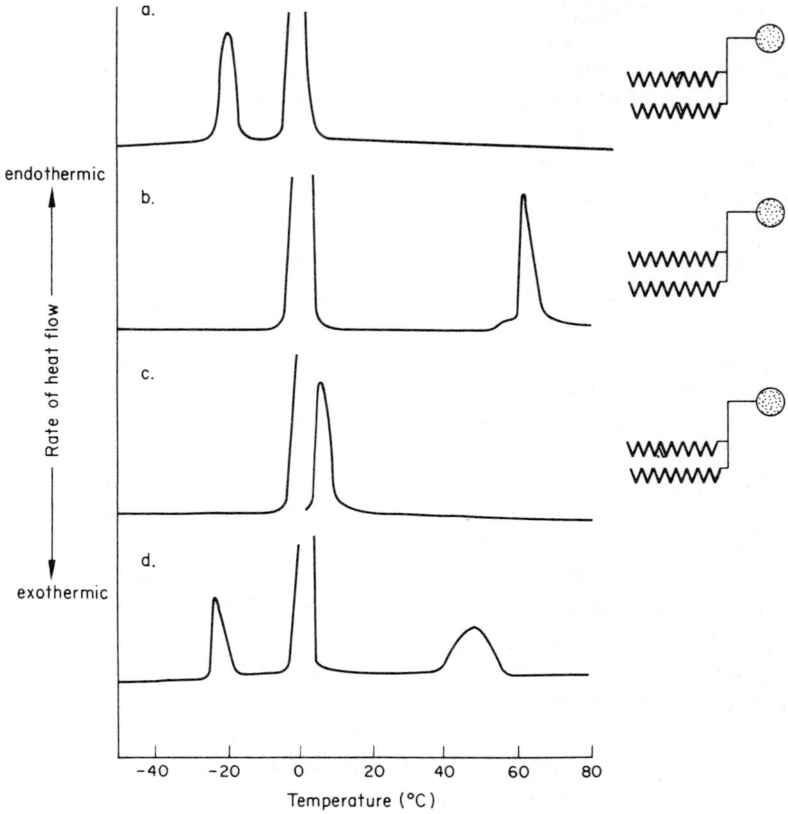

Figure 2.15 Differential scanning calorimeter curves of : (a) dioleoylphosphatidylcholine; (b) distearoylphosphatidylcholine; (c) l-stearoyl-2-oleoylphosphatidylcholine; (d) an equimolar mixture of (a) and (b). All phospholipids were dispersed in an equal mass of water (adapted from reference 43).

T_1) towards the termini of the hydrocarbon chains and towards the quaternary nitrogen atom. The graduation in mobility along the hydrocarbon chains does not appear to conform to either a linear or exponential relationship from one methylene group to the next but indicates a pronounced increase in mobility about the terminal methyl residue in the interior of the bilayer.

2.2.5 Hydrophobic Interactions Between Lipids

The segregation of hydrophobic residues of phospholipids and other membrane constituents from the aqueous environment implies a certain amount of interaction between the hydrocarbon constituents in the interior of the bilayer. Interactions between acyl chains of phospholipids, for example, have been demonstrated by differential scanning calorimetry[43] (figure 2.15). Calorimetry curves of pure 1,2-dioleoylphosphatidylcholine (figure 2.15a) and 1,2-distearoylphosphatidylcholine (figure 2.15b) both exhibit a single endothermic peak at a temperature corresponding to the phase-transition temperature of the particular phospholipid. A single peak is also recorded for 1-steroyl-2-oleoylphosphatidylcholine (figure 2.15c), which appears at a temperature intermediate between those of the phospholipids containing only one type of hydrocarbon chain. When the two pure phospholipids are mixed in equimolar amounts, however, certain changes in the transitions can be discerned, such as a broadening of the transition range, but more significantly there are two separate transitions corresponding to the change in phase of each of the two components of the mixture (see figure 2.15d). It has been suggested that at intermediate temperatures the two molecular species segregate in the bilayer forming domains of crystalline and liquid-crystalline structure. Segregation of phospholipids in bilayers has been confirmed more directly by freeze-fracture techniques. Van Deenen and coworkers[44,45] examined freeze-fracture replicas of liposomes prepared from phospholipids with different hydrocarbon-chain constituents. They found that if certain phospholipids were cooled rapidly (quenched) from a temperature above the phase-transition temperature of that particular phospholipid the fracture face was invariably smooth whereas quenching from below the phase-transition temperature produced a banding effect in the fracture face. The periodicity of these bands appears to be characteristic of the particular phospholipid; for 1-dioleoyl-2-stearoylphosphatidylcholine quenched from − 5°C it was about 50 nm, and for distearoylphosphatidylcholine quenched from 5°C some regions showed regular spacings of 15 nm while others varied from 30 to 50 nm. Dioleoylphosphatidylcholine, however, with a phase-transition temperature of − 20°C, does not exhibit crystallisation structures even if quenched from a temperature of − 40°C. When liposomes prepared from an equimolar mixture of distearoyl- and dioleoylphosphatidylcholines were quenched from an intermediate temperature (5°C), the fracture face was partly covered with a band pattern with a 40 nm periodicity attributed to distearoylphosphatidylcholine interspersed with smooth regions corresponding to dioleoylphosphatidylcholine. These features were retained even if the liposomes were quenched from − 40°C. They concluded that bilayers of mixed phospholipids at a given temperature will form a mosaic with respect to hydrocarbon structure, some regions consisting of chains in liquid-crystalline conformation and the remainder containing molecules in crystalline-hydrocarbon arrangement. It is conceivable that in more complex mixtures phospholipids may be further subdivided on the basis of individual molecular species. The segregation of phospholipids in mixed bilayers implies that the constituent molecules are capable of lateral diffusion in the plane of the bilayer; this has been shown independently by other means and will be examined in more detail later (section 3.2.1).

Undoubtedly the most intensively studied lipid–lipid interaction is that between cholesterol and phospholipids[46]. Nearly fifty years have elapsed since the initial discovery that the addition of cholesterol to phospholipid brings about a reduction in the mean molecular area of the interacting components, creating what is tantamount to a condensed system. The condensing effect has been explained as a specific interaction between cholesterol molecules and liquid-crystal hydrocarbon chains of phospholipids, which hinders the motional freedom of the chains. The effect is only observed in phospholipid mixtures containing disordered hydrocarbon chains and not when these are in a crystalline phase as indicated in figure 2.16. The data were derived from experiments in which monolayers consisting of an equimolar mixture of cholesterol and saturated 1,2-diacylphosphatidylcholines were spread at the air–water interface at 22°C and the mean molecular area at 20 mN m^{-1} was determined. It has been shown previously (see figure 2.7) that these phospholipids are of an expanded type at 20 mN m^{-1} if the hydrocarbon chains are less than sixteen carbon atoms long but that phospholipids with longer chains are condensed. It can be seen that adding cholesterol to condensed phospholipids does not produce any further reduction in mean molecular area.

Hinz and Sturtevant[48] have examined the stoichiometry between phospholipid and cholesterol by differential scanning calorimetry. When increasing amounts of cholesterol were added to an aqueous dispersion of dipalmitoylphosphatidylcholine they found a reduction in the apparent transition enthalpy (ΔH) of the interaction but no shift in the transition temperature. No trace of a thermotrophic phase transition can be observed in the differential scanning calorimeter curves when mole ratio of cholesterol to phospholipid reaches 1:2 and consequently ΔH falls to zero. This suggests that at 33 mole per cent cholesterol each sterol molecule removes effectively four hydrocarbon chains from participation in the co-operative phase transition. A possible arrangement of phospholipid acyl chains and cholesterol molecules in the plane of a mixed bilayer has been proposed by Engelman and Rothman[49] in which each cholesterol molecule is surrounded by seven hydrocarbon chains

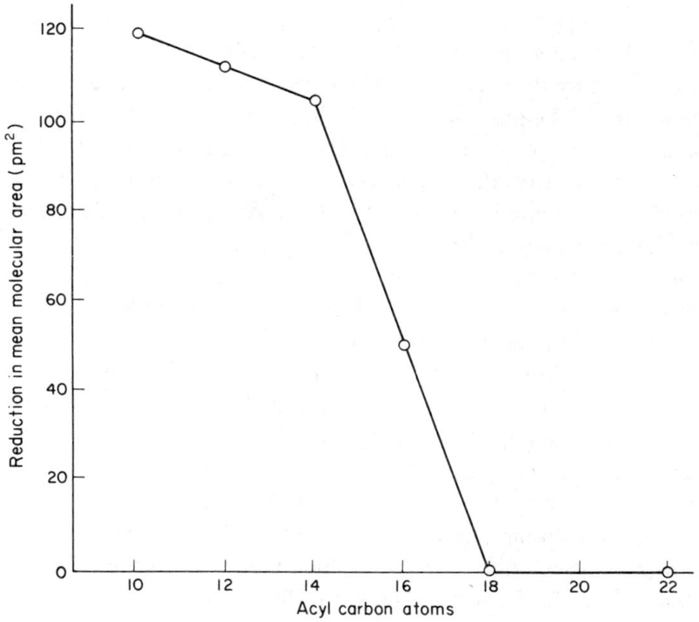

Figure 2.16 Condensing effect of cholesterol on monolayers of a homologous series of saturated 1,2-diacylphosphatidylcholines. The reduction in mean molecular area from that of an ideally mixed system of equimolar amounts of each lipid is plotted as a function of hydrocarbon chain length (data from reference 47).

giving an overall stoichiometry of about two phospholipid molecules per cholesterol.

Additional information regarding the packing arrangement of cholesterol in phospholipid bilayers has been obtained from spin-label experiments with paramagnetic steroid analogues incorporated into dipalmitoylphosphatidylcholine dispersions[25]. At temperatures above the phase-transition temperature (about 41°C) the probe appears to be evenly distributed among the mobile hydrocarbon chains indicating an ideal mixing of the components. When the temperature is reduced below the phase-transition temperature, however, a mosaic structure develops in which small clusters of steroid molecules segregate within the lipid matrix. On reheating, the clusters disappear and the lipids are again ideally mixed, suggesting that the process is completely reversible and that the lipids are capable of translational mobility in the plane of the bilayer. The spin-label steroid analogue used in these experiments differs in a number of respects from cholesterol but the separation of cholesterol and phospholipid phases in mixed monolayers below the phase-transition temperature has been confirmed in monolayer experiments. Pagano and Gershfeld[50] used a surface vapour-pressure technique by which equilibrium conditions can be readily established to show that, at 25°C, dipalmitoylphosphatidylcholine and cholesterol were not ideally mixed in monomolecular films.

2.2.6 Electrostatic Properties of Charged Lipids

The stability of lipid bilayers in aqueous dispersion depends both on the segregation of hydrocarbon residues from the aqueous phase and on a polar interaction of the lipid with water. The polar interaction of most membrane lipids is dominated by the presence of a charged group, which constitutes the polar 'headgroup' of the molecule. Phospholipids, for example, bear a partial negative charge on the phosphate group, which dissociates at about pH 2–3. The structure and electrostatic properties of the polar group of various phospholipids are presented in table 2.2. Other charged groups such as sulphate (cerebroside sulphate) and carboxyl groups (phosphatidylserine) are also present on some membrane lipids. Apart from electrostatic interactions, strong polar interactions arise from carbohydrates attached to certain lipids such as phosphatidylinositol and the glycolipids. In this case the dipoles arise from the numerous hydroxyl groups of the sugar residues. The single hydroxyl group of cholesterol confers a weak amphipathic character to the lipid, which enables it to orientate specifically in phospholipid structures.

When charged phospholipids are oriented, as for example in a monomolecular film or a bilayer structure, the charged groups will reside in a plane near the junction of the aqueous and hydrophobic regions. The negative potential generated by the dissociation of the phosphate groups along this plane, designated Ψ_0, depends on the density and degree of dissociation of the ionisable groups, and the potential decreases exponentially with distance from the plane. The presence of similarly charged groups on adjacent molecules sets up an electrostatic repulsion and results in an appreciable expansion in the area occupied by each molecule in the plane of the structure. In bilayers of charged

Table 2.2 The structure of phospholipid charged groups and their electrostatic properties

Phospholipid ionisable groups	Structure	Electrostatic properties
1. primary phosphate (phosphatidic acid)	$R-P(=O)(O^-)-O^-$	pK_1 3.9 pK_2 8.3
2. secondary phosphate (phosphatidylinositol cardiolipin)	$R-P(=O)(O^-)-O-R$	$pK < 2.0$
3. secondary phosphate + quaternary amine (sphingomyelin phosphatidylcholine)	$R-P(=O)(O^-)-O-CH_2-CH_2-{}^+N(CH_3)_3$	isoelectric in pH range 3 to 10
4. secondary phosphate + amine (phosphatidylethanolamine)	$R-P(=O)(O^-)-O-CH_2-CH_2-{}^+NH_3$	net negative at pH 7.4
5. secondary phosphate + amine + carboxyl (phosphatidylserine)	$R-P(=O)(O^-)-O-CH_2-CH({}^+NH_3)-C(=O)O^-$	net negative at pH 7.4

lipids, the electrostatic repulsion between similarly charged planes on either side of the bilayer prevents a thinning of the structure and supplements the lateral force of cohesion between hydrocarbon chains, thus stabilising the two layers. The charged plane will attract oppositely charged mobile counterions from the surrounding aqueous phase (sometimes referred to as gegenions) and at equilibrium they will be distributed, according to the electrostatic potential, to form a so-called electrical double layer. In the absence of specific interactions between the charged groups of the lipid and the counter ions (that is, only charge–dipole interactions) the decay in potential away from the charged plane will approximate to a Boltzmann distribution as illustrated in figure 2.17a. Phospholipid dispersions consisting of charged lipids will migrate towards the anode if placed in a potential gradient and in dilute salt solutions the particles will be accompanied by a 'cloud' of immobile counter ions, which are separated along a plane of shear from the more remote mobile ions. The electrical potential at the plane of shear is referred to as the zeta (ζ) potential.

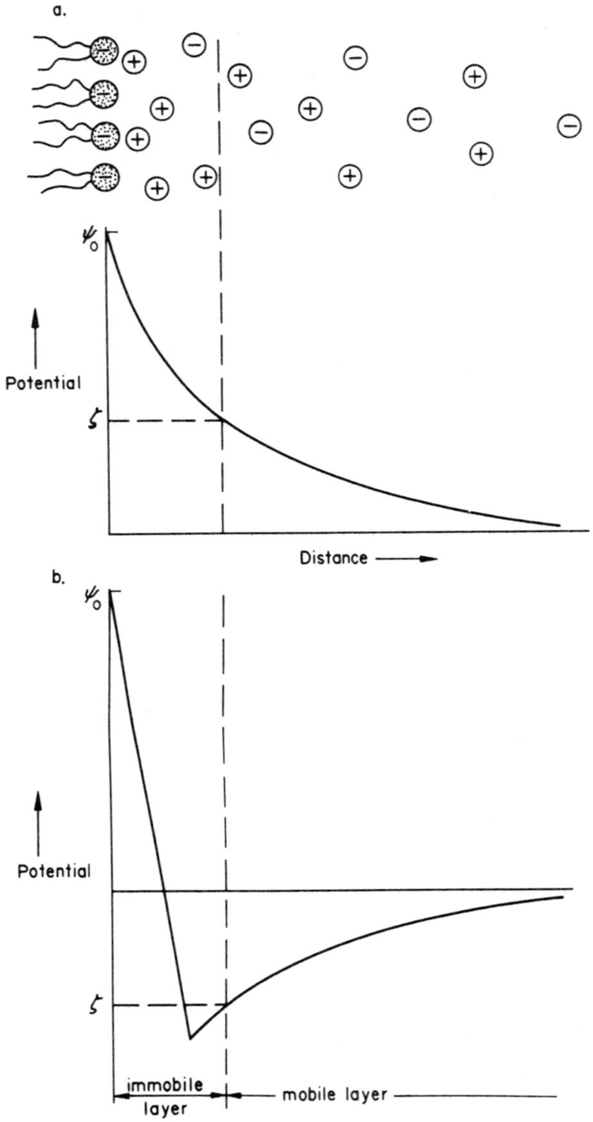

Figure 2.17 The variation in potential with distance from a charged phospholipid surface. The counterion distribution in an electrical double layer with no specific interactions between the molecules is illustrated schematically (a) and consists of an immobile layer close to the charged surface and a mobile layer further removed. (b) Potential variation with a strong specific interaction between charged groups. Note that the ζ potential is reversed (adapted from reference 51).

Certain counter ions such as $(UO_2)^{2+}$ interact more specifically with ionised groups of lipids so that in addition to electrostatic attractions there may be van der Waals and other dipole–dipole forces, promoting a more intimate association between the charged components. Where these forces predominate it is possible to reverse the zeta potential, and phospholipid dispersions will then migrate towards the cathode in an applied potential gradient (see figure 2.17b). Zwitterionic phospholipids such as phosphatidylcholine and sphingomyelin possess a positively charged quaternary amino group in addition to the negatively charged phosphate group, and the ionised groups are believed to interact to form an internal salt linkage. Whether or not these phospholipids can attract counterions exclusively through charge–dipole interactions is debatable but, if they do, the electrostatic component is likely to be extremely weak. Other types of interaction between phosphatidylcholine dispersions and ions such as $(UO_2)^{2+}$ are indicated by the fact that these ions cause a shift in the phase-transition temperature of dipalmitoylphosphatidylcholine from 41.5°C to about 46°C[19].

In an analysis of calcium binding to monomolecular films of cerebroside sulphate, Quinn and Sherman[52] concluded that the binding site consisted of two adjacent lipid molecules bridged by the hydrated divalent ion situated close to the charged plane of the film. Further evidence of divalent bridging of phospholipids by calcium has been provided by Ohnishi and Ito[53] using spin-label methods. They incorporated a spin-label analogue of phosphatidylcholine into a mixture of anionic phosphatidylserine and phosphatidylcholine, and impregnated the lipids into millipore filters. Addition of calcium ions produced a rapid and reversible (with calcium sequestering agents) exchange-broadening of the electron spin resonance signal, which they attributed to a two-dimensional phase separation of the phospholipids. They suggested that phosphatidylserine was clustered by divalent bridging with calcium ions leaving a separate fluid phase of phosphatidylcholine.

Another feature of the negatively-charged surface of phospholipid structures is that protons derived from the dissociation of water molecules are attracted to the surface layer, causing a decrease in pH of the surface phase (pH_s) relative to the bulk aqueous phase (pH_b). The increase in proton concentration in the surface is related to surface potential by

$$pH_s = pH_b + \frac{e\Psi_o}{2.3kT} \qquad (2.7)$$

where e is the electronic charge; Ψ_o is the surface potential (in mV) of the charged plane; k is the Boltzmann constant and T, the absolute temperature. It can be seen from equation 2.7 that the apparent pK of ionisation of surface-charged groups will be somewhat higher than the corresponding groups in free solution, and in the case of phosphate groups of phospholipids this may be one or two pH units more alkaline, depending on the surface-charge density. Surface-potential measurements of charged monomolecular films at the air–water interface indicate that protons (presumably in the form of hydronium ions, H_3O^+) continue to interact with the polar groups of completely ionised lipids up to pH 7 (where $[H_3O^+] = [OH^-]$), probably through hydrogen bonds or other dipole–dipole forces[54]. Monolayers of lipids with no ionisable groups, such as cerebroside, also show large changes in surface potential centred about pH 6.5, which are believed to reflect changes in water structure about the hydrated sugar residue[52].

2.3 The Aqueous Phase

Water is required to orientate amphipathic lipid molecules into bilayers, and consequently the polar character of the aqueous phase largely determines the intrinsic structure of membranes. In the absence of water, amphipathic lipids exist in different molecular arrangements[17], although some physical features are common to lipids in different structural conformations. Anhydrous distearoylphosphatidylcholine, for example, undergoes an endothermic phase transition but the transition tends to be nonco-operative and occurs at a much higher temperature[55] (approximately 85°C). Water appears to 'loosen' the structure and allows the hydrocarbon chains to become more flexible, because the addition of water to the anhydrous phospholipid sharpens the endothermic reaction and decreases the phase-transition temperature to 58°C when in excess (about 40 per cent by weight; see figure 2.5). The amount of bound water, taken as that water which does not participate in the ice-melting peak at 0°C, is found to be about ten molecules per polar headgroup.

Rather surprisingly, both phospholipid bilayers and biological membranes are relatively permeable to water compared with other small polar molecules (see section 4.1.1). This can be demonstrated simply by observing changes in volume (usually monitored by light-scattering techniques) when impermeative solutes are added to liposome dispersions. Such experiments have shown that charged phospholipid bilayers behave as almost perfect osmometers. Huang and Thompson[56] have measured the water permeativity coefficients of artificial lipid bilayer membranes both by diffusion exchange of isotopically labelled water and by measuring the volume of the net water flux across the membrane. Bilayer membranes can be formed by applying lipid dissolved in a suitable solvent to a small orifice in a septum dividing two aqueous compartments. After the excess lipid and solvent have drained away a bilayer of lipid remains stretched across the hole. A more recent method, shown diagrammatically in

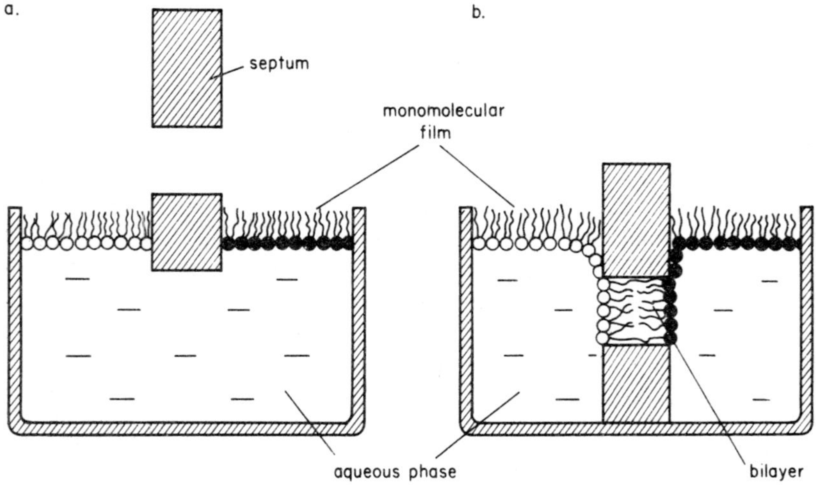

Figure 2.18 Schematic representation of a method for producing an asymmetric lipid bilayer separating two aqueous compartments.

figure 2.18, involves spreading two monolayers of amphipathic lipids on each surface divided by a raised septum (figure 2.18a); the composition of each monolayer can obviously be adjusted to produce asymmetric bilayers. The septum is then carefully lowered into position so that it divides the two aqueous compartments and during this process a bilayer forms as the orifice passes through the surface films (figure 2.18b). The data presented in table 2.3 show, by a number of physical criteria, that these membranes bear a remarkable resemblance to biological membranes. Isotope exchange of ^3H_2O through bilayers of egg-yolk phosphatidylcholine has been determined by measuring the rate at which isotopically labelled water passes from one compartment to the other. Huang and Thompson[56] calculated the permeativity coefficient of ^3H_2O through these bilayers at 36°C and obtained a value of 4.4 μm s^{-1}, corresponding to a flux of nearly 3×10^{24} molecules of water penetrating each cm^2 of bilayer per second. Osmotic flux measurements were performed by recording the rate of volume decrease of one compartment separated by the bilayer when the osmolality of the other compartment was increased. Flux rates varying between 17 and 104 μm s^{-1} were obtained and no satisfactory explanation could be offered to account for the discrepancy between the two methods although the rate appeared to depend markedly on the particular phospholipid preparation used to form the bilayer.

Table 2.3 The physical characteristics of artificial bilayers and biological membranes

	Phospholipid bilayers	Biological membranes
*thickness (nm)	4.5–10	4–12
interfacial tension (kJ m^{-2})	2–60	0.3–30
refractive index	1.56–1.66	approx. 1.6
electrical resistance (Ωm^{-2})	10^7–10^{13}	10^6 - 10^9
capacitance (mF m^{-2})	3–13	5 - 13
breakdown voltage (mV)	100–550	100
resting potential difference (mV)	0–140	10–88

*These measurements have been made by electron microscopy, X-ray diffraction and optical methods. The capacitance was determined using an assumed dielectric constant.

Numerous attempts have been made to assess the extent to which the surface film influences the underlying water structure. Most of these experiments have been conducted by measuring the volume of water carried along with monomolecular films either flowing along a surface–pressure gradient or transposed between barriers at constant area per molecule across an aqueous substrate. Crisp[57] has calculated that water to a depth of 2500 nm moves with monolayers of oleic acid, a value more than three orders of magnitude greater than the thickness of a bilayer. Fatty acid monolayers of surface density 0.3 nm^2 per molecule carry a volume of water equivalent to 300 000 molecules per molecule of fatty acid, a figure that far exceeds the possibility of all water molecules interacting intrinsically with the film. It does seem likely, however, that some long-range order exists in water molecules adjacent to the surface, but more precise information will be required before the extent of this order can be firmly established.

2.4 Summary

The protein and lipid components of biological membranes associate together by nonbonding forces, and the manner in which they interact depends on the amphipathic character of the individual constituents. Most membrane lipids possess hydrophobic residues in the form of long hydrocarbon chains which are separated within the molecule from a polar region consisting of one or more ionised groups or, in certain lipids, one or more carbohydrate residues. The amphipathic character of membrane proteins depends on the number and relative disposition of amino acid side chains in the folded peptide chain. The solubility of intrinsic membrane proteins in solvents of differing polarity, and their tendency to interact with hydrophobic residues of membrane lipids, indicates that they are amphipathic. Moreover, such proteins contain a higher proportion of amino acids with hydrophobic side chains than found in typically water-soluble proteins.

Spectroscopic methods are used to determine the average secondary structure of membrane proteins *in situ*. These include infrared absorption and measurement of optical activity in ultraviolet light. In general, the predominant coiling arrangement of the peptide chain of membrane proteins is α-helix or random coil and, with the possible exception of the inner mitochondrial membrane, β-conformation is absent. Estimates of α-helical content vary from one membrane to another but most membranes contain proteins with between 30 and 50 per cent α-helix, with the remainder in random-coil arrangement. The proportion of α-helix may reflect the nature and scope of enzymic processes associated with particular membranes.

Membrane lipid extracts and purified phospholipid fractions form lamellar structures on gentle dispersion in aqueous solvents. These structures, referred to as liposomes, consist of stacked arrays of bimolecular leaflets of lipid with intervening layers of water. Vigorous dispersion such as irradiation with high-energy ultrasound breaks down the liposomes into small single bilayer vesicles. X-ray diffraction studies indicate that the molecules are orientated with the hydrocarbon chains extending into the interior of the bilayer and the polar headgroup exposed to water on the outside; this arrangement is particularly favourable from a thermodynamic viewpoint.

The hydrocarbon chains of membrane phospholipids can exist in two temperature-dependent forms, an ordered crystalline lattice at lower temperatures and a disordered liquid-crystal phase at higher temperatures. The particular phase at any given temperature is a function of the length and degree of saturation of the hydrocarbon chains and depends to some extent on the nature of the polar interaction with water. The transition from a crystal to a liquid-crystal phase is associated with an increase in volume and lateral expansion of the bilayer owing to the increased motion of the hydrocarbon chains. This mobility has been examined by fluorescent-probe techniques and nuclear magnetic resonance spectroscopy, and it has been shown that the motion of the hydrocarbon chains increases with distance from the surface of the bilayer; the terminal methyl residues are by far the most mobile. Studies with phospholipid mixtures suggest that these molecules segregate in the plane of the bilayer according to the molecular arrangement of the hydrocarbon chains. More specific interactions between cholesterol and expanded phospholipids cause a restriction in hydrocarbon chain motion.

Ionisable groups of membrane lipid molecules introduce a negative surface charge on to lipid structures. The surface potential depends on the surface density of the ionisable groups and the potential decays with distance away from the surface. The presence of a surface charge has two main effects on lipid structures: (1) electrostatic repulsion between similarly charged planes on either side of the bilayer prevents thinning and collapse of the structure; (2) counter-ions of opposite charge are attracted to the surface and are distributed in the aqueous phase in the form of an electrical double layer. The layers consist of an inner immobile layer, which accompanies the migration of the charged plane when placed in a potential gradient, and an outer mobile layer, which does not. The potential at the plane of shear between the two layers (zeta potential) is always the same sign as the surface potential when only electrostatic forces are involved, but the sign of the zeta potential may be reversed with more specific interactions. The negative surface potential also attracts protons to the surface causing a decrease in surface pH and an apparent increase in the pK of ionisation of the charged groups. Charged groups and other dipole residues are thought to exert an order among the surrounding water molecules and it is likely that such effects may extend for some considerable distance from the surface. Lipid bilayers appear to be relatively permeable to water.

References

1. R.E. Dickerson and I. Geis, *The Structure and Action of Proteins,* Harper & Row, New York (1969)
2. R.A. Capaldi and G. Vanderkooi. The low polarity of many membrane proteins. *Proc. natn. Acad. Sci. U.S.A.,* **69** (1972), 930–2
3. D. Chapman, V.B. Kamat and R.J. Levene. Infrared spectra and the chain organization of erythrocyte membranes. *Science, N.Y.,* **160** (1968), 314–16
4. T.J. Jenkinson, V.B. Kamat and D. Chapman. Physical studies of myelin II. Proton magnetic resonance and infrared spectroscopy. *Biochim. Biophys. Acta,* **183** (1969), 427–33
5. J.M. Steim. Spectroscopic and calorimetric studies of biological membrane structure. *Adv. Chem. Ser.,* **84** (1968), 259–302
6. G. Holzwarth. Ultraviolet spectroscopy of biological membranes. In: *Membrane Molecular Biology* (eds. C.F. Fox and A.D. Keith,), Sinauer, Stamford, Conn. (1972), pp 228–86
7. I. Tinoco and C.R. Cantor. Application of optical rotatory dispersion and

circular dichroism to the study of biopolymers. *Meth. biochem. Analysis,* **18** (1970), 81-203

8. L. Masotti, G. Lenaz, A. Spisini and D.W. Urry. Effect of phospholipids on the protein conformation in the inner mitochondrial membranes. *Biochem. biophys. Res. Commun.,* **56** (1974), 892-7
9. D.W. Urry. Conformation of protein in biological membranes and a model transmembrane channel. *Annal. N.Y. Acad. Sci.,* **195** (1972), 108-25
10. G.G. Shipley. Recent X-ray diffraction studies of biological membranes and membrane components. In: *Biological Membranes,* Vol. 2, (D. Chapman and D.F.H. Wallach eds), Academic Press, London (1973), pp 1-89
11. M.H.F. Wilkins, A.E. Blaurock and D.M. Engelman. Bilayer structure in membranes. *Nature new Biol.* **230** (1971), 72-6
12. Y.K. Levine and M.H.F. Wilkins. Structure of oriented lipid bilayers. *Nature new Biol.,* **230** (1971), 69-72
13. C. Mettenheimer. Corr.-Blatt des Vereins für gemeinsame *Arbeit zur Förderung der wissensch. Heilkunde* No. **24** (1857), 331
14. W.H. Wynn. The minute structure of the medullary sheath of nerve-fibres. *J. Anat. Physiol., Lond.* **34** (1900), 381-97
15. W.J. Schmidt. Doppelbrechung und Feinbau der Markscheide der Nervenfasern. *Z. Zellforsch. mikrosk. Anat.,* **23** (1936), 657-76
16. V. Luzzati, T. Gulik-Krzywicki, E. Rivas, F. Reiss-Husson and R.P. Rand. X-ray study of model systems: structure of the lipid-water phases in correlation with the chemical composition of the lipids. *J. gen. Physiol.,* **51** (1968), 37_s-43_s
17. T. Gulik-Krzywicki, E. Rivas and V. Luzzati. Structure et polymorphisme des lipides: etude par diffraction des rayons X du systeme formé de lipides de mitochondries de coeur de boeuf et d'eeu. *J. molec. Biol.,* **27** (1967), 303-22
18. D. Chapman. An infrared spectroscopic examination of some anhydrous sodium soaps. *J. chem. Soc. §,* **152** (1958), 784-9
19. D. Chapman. Some recent studies of lipids, lipid-cholesterol and membrane systems. In: *Biological Membranes,* Vol. 2 (D. Chapman and D.F.H. Wallach, eds.), Academic Press, London (1973), pp 91-144
20. D. Chapman, R.M. Williams and B.D. Ladbrooke. Physical studies of phospholipids. VI. Thermotropic and lyotropic mesomorphism of some 1,2-diacyl phosphatidylcholines (lecithins). *Chem. Phys. Lipids,* **1** (1967), 445-75
21. B.D. Ladbrooke and D. Chapman. Thermal analysis of lipids, proteins and biological membranes. A review and summary of some recent studies. *Chem. Phys. Lipids,* **3** (1969), 304-67
22. H. Traüble and D.H. Haynes. The volume change in lipid bilayer lamellae at the crystalline-liquid crystalline phase transition. *Chem. Phys. Lipids,* **7** (1971), 324-35

23. M.C. Phillips and D. Chapman. Monolayer characteristics of saturated 1,2-diacylphosphatidylcholines (lecithins) and phosphatidylethanolamines at the air–water interface. *Biochim. Biophys. Acta,* **163** (1968), 301–13
24. J.A.G. Taylor, J. Mingins, B.A. Pethica, B.Y.J. Tan and C.M. Jackson. Phase changes and mosaic formation in single and mixed phospholipid monolayers at the oil–water interface. *Biochim. Biophys. Acta,* **323** (1973), 157–60
25. E. Sackmann and H. Traüble. Studies of the crystalline–liquid crystalline phase transition of lipid model membranes I. Use of spin labels and optical probes as indicators of the phase transition. II Analysis of electron spin resonance spectra of steroid labels incorporated into lipid membranes. III Structure of a steroid–lecithin system below and above the lipid-phase transition. *J. Am. chem. Soc.,* **94** (1972), 4482–510
26. M. Shinitzky, A.-C. Dianoux, C. Gitler and G. Weber. Microviscosity and order in the hydrocarbon region of micelles and membranes determined with fluorescent probes. I. Synthetic micelles. *Biochemistry,* **10** (1971), 2106–13
27. U. Cogan, M. Shinitzky, G. Weber and T. Nishida. Microviscosity and order in the hydrocarbon region of phospholipid and phospholipid–cholesterol dispersions determined with fluorescent probes. *Biochemistry,* **12** (1973), 521–8
28. P.J. Quinn and Y. Barenholz. A comparison of the activity of phosphatidylinositol phosphodiesterase against substrate in dispersions and as monolayers at the air–water interface. *Biochem. J.* **149** (1975), 199–208
29. K.A. McLauchlan, *Magnetic Resonance,* Clarendon Press, Oxford (1972)
30. I.C.P. Smith. The spin label method. In: *Biological Applications of Electron Spin Resonance* (H.M. Swartz, J.R. Bolton and D.C. Borg, eds), Wiley–Interscience, New York (1972), pp 483–539
31. R.J. Mehlhorn and A.D. Keith. Spin labeling of biological membranes. In: *Membrane Molecular Biology* (C.F. Fox and A.D. Keith, eds) Sinauer, Stamford, Conn. (1972), pp 192–227
32. W.L. Hubbell and H.M. McConnell. Molecular motion in spin-labeled phospholipids and membranes. *J. Am. chem. Soc.,* **93** (1971), 314–26
33. A. Carrington and A.D. McLachlan, *Introduction to Magnetic Resonance with Applications to Chemistry and Chemical Physics,* Harper & Row, New York (1967)
34. J.D. Roberts, *Nuclear Magnetic Resonance, Applications to Organic Chemistry,* McGraw-Hill, New York (1959)
35. A.F. Horwitz. Nuclear magnetic resonance studies on phospholipids and membranes. In: *Membrane Molecular Biology* (C.F. Fox and A.D. Keith, eds), Sinauer, Stamford, Conn. (1972), pp 164–91
36. D. Chapman and A. Morrison. Physical studies of phospholipids. IV High resolution nuclear magnetic resonance spectra of phospholipids and related

substances. *J. biol. Chem.*, **241** (1966), 5044–52

37. D. Chapman and S.A. Penkett. Nuclear magnetic resonance spectroscopic studies of the interaction of phospholipids with cholesterol. *Nature, Lond.*, **211** (1966), 1304–5
38. B. Sheard. Internal mobility of phospholipids. *Nature, Lond.*, **223** (1969), 1057–9
39. A.F. Horwitz, D. Michaelson and M.P. Klein. Magnetic resonance studies on membrane and model membrane systems. III fatty acid motions in aqueous lecithin dispersions. *Biochim. Biophys. Acta,* **298** (1973), 1–7
40. M. Hsu and S.I. Chan. Nuclear magnetic resonance studies of the interaction of valinomycin with unsonicated lecithin bilayers. *Biochemistry,* **12** (1973), 3872–6
41. A.G. Lee, N.J.M. Birdsall, Y.K. Levine and J.C. Metcalfe. High resolution proton relaxation studies of lecithins. *Biochim. Biophys. Acta,* **255** (1972), 43–56
42. Y.K. Levine, N.J.M. Birdsall, A.G. Lee and J.C. Metcalfe. ^{13}C nuclear magnetic resonance relaxation measurements of synthetic lecithins and the effect of spin-labelled lipids. *Biochemistry,* **11** (1972), 1416–21
43. M.C. Phillips, H. Hauser and F. Paltauf. The inter- and intra-molecular mixing of hydrocarbon chains in lecithin/water systems. *Chem. Phys. Lipids,* **8** (1972), 127–33
44. A.J. Verkleij, P.H.J.Th. Ververgaert, L.L.M. Van Deenen and P.F. Elbers. Phase transitions of phospholipid bilayers and membranes of *Acholeplasma Laidlawii B* visualized by freeze fracturing electron microscopy. *Biochim. Biophys. Acta,* **288** (1972), 326–32
45. P.H.J.Th. Ververgaert, A.J. Verkleij, P.F. Elbers and L.L.M. Van Deenen. Analysis of the crystallization process in lecithin liposomes: a freeze-etch study. *Biochim. Biophys. Acta,* **311** (1973), 320–9
46. M.C. Phillips. The physical state of phospholipids and cholesterol in monolayers, bilayers and membranes. In: *Progress in Surface and Membrane Science,* Vol. 5, (J.F. Danielli, M.D. Rosenberg and D.A. Cadenhead, eds), Academic Press, New York (1972), pp 139–221
47. D. Chapman, N.F. Owns, M.C. Phillips and D.A. Walker. Mixed monolayers of phospholipids and cholesterol. *Biochim. Biophys. Acta,* **183** (1969), 458–65
48. H-J. Hinz and J.M. Sturtevant. Calorimetric investigation of the influence of cholesterol on the transition properties of bilayers formed from synthetic L-α-lecithins in aqueous suspensions. *J. biol. Chem.*, **247** (1972), 3697–700
49. D.M. Engelman and J.E. Rothman. The planar organization of lecithin-cholesterol bilayers. *J. biol. Chem.*, **247** (1972), 3694–7
50. R.E. Pagano and N.L. Gershfeld. Phase separation in cholesterol-dipalmitoyl lecithin mixed films and the condensing effect. *J. Colloid Sci.*, **44** (1973), 382–3

51. J.T. Davies and E.K. Rideal, *Interfacial Phenomena,* Academic Press, New York (1961), p 88

52. P.J. Quinn and W.R. Sherman. Monolayer characteristics and calcium adsorption to cerebroside and cerebroside sulphate oriented at the air–water interface. *Biochim. Biophys. Acta,* **233** (1971), 734–52

53. S. Ohnishi and T. Ito. Calcium-induced phase separations in phosphatidylserine–phosphatidylcholine membranes. *Biochemistry*, **13** (1974), 881–7

54. P.J. Quinn and R.M.C. Dawson. The pH dependence of calcium adsorption onto anionic phospholipid monolayers. *Chem. Phys. Lipids,* **8** (1972), 1–9

55. R.M. Williams and D. Chapman. Phospholipids, liquid crystals and cell membranes. *Prog. Chem. Fats, Lipids,* **11** (1971), 3–79

56. C. Huang and T.E. Thompson. Properties of lipid bilayer membranes separating two aqueous phases: water permeability. *J. Molec. Biol.,* **15** (1966), 539–54

57. D.J. Crisp. Two-dimensional transport at fluid interfaces. *Trans. Faraday Soc.,* **42** (1946), 619–35

3 Molecular Organisation in Membranes

3.1 Models of Membrane Structure

The molecular arrangement in cell membranes was originally conceived as a static array of metabolically inert molecules. Membrane models constructed on this principle were able to provide a satisfactory account of certain properties of cell membranes, such as the permeability of some molecules but not others across the membrane, and were consistent with early microscopic studies of membrane anatomy. Contemporary views of membrane structure are, however, profoundly different, and have been shaped largely from information derived from a better understanding of the molecular microenvironment within membranes. Membranes are now depicted as a dynamic array of molecules, highly mobile within the structure and constantly interacting with other molecules within the membrane as well as in the surrounding media. Many processes conducted in and around membranes appear to depend on specific interactions between different components of the membrane. This can be seen in, for example, the loss of activity of certain membrane-bound enzymes when phospholipids are removed; full activity can often be restored by putting back specific phospholipids in the depleted membrane. It is apparent therefore that detailed appreciation of molecular interactions between membrane components will be required before an accurate formulation of the structure of any particular membrane can be attempted.

Membrane models have, in the past, been constructed from the vast amount of information obtained from plasma membranes of the myelin sheath and retinal-rod outer segment membranes, for which accurate physical measurements can be obtained. Both of these membranes, however, have relatively low amounts of protein compared to the lipid fraction, which tends to dominate their structure. It can be cogently argued that models of membrane structure based on analyses of myelin and retinal-rod outer-segment membranes may not be appropriate for those, like the inner mitochondrial membrane, where there is substantially more protein relative to lipid. We might ask, therefore, whether a single membrane model can be formulated to embrace all cell membranes, or whether the structure of each particular membrane is in some way unique and distinct from all others. The answer probably lies somewhere between these two views, since there is likely to be some degree of similarity between different membranes, while each membrane may differ in detail depending on its molecular composition and on the manner in which the components interact. We shall now address ourselves to an examination of those features that appear to be common to cell membranes in general and discuss, where appropriate, some of the distinctive features of particular cell membranes.

3.1.1 Lipid Bilayer in Cell Membranes

Many earlier models of membrane structure incorporated a lipid bilayer as the salient structural feature. The provocative evidence for these models was obtained from the experiments of Gorter and Grendel[1], who began with the premise that membranes were, in part, lipoidal in character, since hydrophobic molecules were known to penetrate membranes more easily than hydrophilic ones[2]. Furthermore, membrane lipids were shown to orientate themselves in a predictable manner at oil-water or air-water interfaces[3]. They knew that lipids were an important structural component of cell membranes but had no idea of how they were organised within the membrane. To answer this question they extracted lipids from erythrocytes with acetone and compared the area occupied by the lipids when orientated as a monomolecular film at the air-water interface with the area of membrane from which they were extracted. The area covered by the film was almost twice that of the membrane area, so they concluded that the membrane consisted of a two-dimensional array of lipids, two molecules thick. The original experiments of Gorter and Grendel have since been scrutinised by Bar *et al.*[4], who reported that although, experimentally, the surface-area approach appeared to be sound there were certain technical errors in their method. First, acetone extraction is likely to remove only 70–80 per cent of total membrane lipid. In the original experiments this incomplete lipid extraction was compensated for by under-estimation of the surface area of the red-cell membrane. A value of 99 μm^2/red-cell area was used, obtained from dehydrated samples. This is somewhat less than the best contemporary estimates of red-cell area, which are in the region of 145 μm^2 for hydrated specimens. The second point raised by Bar and his co-workers concerned the surface pressure at which the film areas were measured. Since there is still some uncertainty about the precise packing density of lipid molecules into bilayers, this parameter remains conjectural although X-ray and other measurements (see section 2.2.2) indicate that fairly dense packing arrangements are likely. Nevertheless, when the total lipid extract of erythrocyte membranes is spread at the air–water interface, the minimum area occupied by the film is equivalent to about 65 per cent of the total membrane area, assuming a bimolecular arrangement. Similar calculations have now been extended to other cell membranes for which the necessary chemical data are available (see table 3.1), and in general the proportion of membrane area that could exist as a lipid bilayer is inversely related to the protein content of the membrane. These calculations assume that lipid molecules are packed tightly together in the membrane, occupy an area equivalent to the limiting area obtained at an air–water interface and are associated with about 20 per cent by weight of water in the membrane. There is reasonable agreement with previous measurements of the surface area of erythrocyte membranes covered by lipid bilayer. This is true even if slightly expanded films or different water contents are substituted in these calculations so that no definitive conclusions can be drawn with regard to lipid packing density or the extent to which the membrane is hydrated.

A different approach to calculating the bilayer area of erythrocyte membranes has been attempted by Engelman[5], in which he has avoided assumptions concerning the degree of compression of monomolecular films when measuring

Table 3.1 The proportion of membrane surface area consisting of lipid bilayer, calculated from the limiting area occupied by a total membrane lipid extract orientated as a monomolecular film at an air–water interface.

Membrane	Ratio protein: total lipid (wt./wt.)	Maximum per cent surface area accounted for as bilayer
plasma membrane		
myelin	0.28	103
erythrocyte	1.50	67
acholeplasma laidlawii	1.78	62
endoplasmic reticulum	0.90	83
sarcoplasmic reticulum	1.00	80
mitochondrial membranes		
inner	3.55	40
outer	1.22	72

film areas. First, he calculated the volume of membrane occupied by the hydrocarbon residues of phospholipid molecules, assigning an average length of 16.5 carbon atoms to acyl hydrocarbon chains extending into the hydrophobic region, and a mean of 1.26 unsaturated bonds per chain. To this he added the volume of cholesterol, obtained from density measurements, assuming that the entire molecule resides in the hydrophobic region of the membrane. The value obtained was too large for a membrane consisting of a single layer of lipid molecules, yet insufficient for one covering the entire red-cell surface in bilayer configuration. He concluded that the remaining 10–20 per cent of the hydrophobic region of the membrane consisted of nonlipid components.

A more direct method of estimating the surface area of the plasma membrane that is occupied by phospholipids has been to remove these lipids selectively and then to correlate the results with the reduction in membrane surface area. Red-cell membrane preparations, for example, have been treated with phospholipase c to convert phospholipids into diglycerides and water-soluble products[6]. Approximately 70 per cent of the total membrane phospholipids were found to be suscepitble to hydrolysis by phospholipase C and the diglyceride product accumulates in the form of small droplets, which remain associated with the membrane. When digestion of susceptible phospholipids was complete, the surface area of the membrane was reduced to about half the original area and no apparent morphological changes were observed when thin sections of the enzyme-treated membrane were examined under the electron microscope. Freeze–cleave electron microscopy of partially digested membrane preparations showed a direct correlation between the density of membrane-associated particles visible in the inner membrane fracture faces and the amount of phospholipid hydrolysed, until about half the total phospholipid had been degraded. Correlation with more extensively digested membranes was not possible because few membrane cleavage planes were produced in these preparations. Furthermore, these results confirm that about 70 per cent of the erythrocyte membrane area consists of lipid, and they are consistent with the supposition that membrane-associated particles are unlikely to consist of phospholipid, since the number of particles remained constant after phospholipase treatment. The overall conclusion from these experiments is that a bilayer arrangement of lipids is probably an important structural feature of most cell membranes.

3.1.2 Integration of Protein into Membranes

The next, and possibly more difficult, problem is to decide how membrane proteins are integrated into the lipid bilayer. Many of the earlier models of membrane structure depicted proteins as being peripheral to the underlying lipid bilayer. Among the first seriously to contemplate protein as a membrane component were Danielli and Davson[7]: they proposed a model in which globular proteins were shown attached to the polar phospholipid headgroups on each side of the bilayer (figure 3.1a). This model was subsequently modified by Robertson[8] in the light of new information, especially concerning the ultra-structural features of membranes that he was able to observe under the electron microscope. Most of these studies were of plasma membranes of myelin but the membrane model proposed (figure 3.1b), usually referred to as the unit-membrane hypothesis, was suggested as the basic structure of all cell membranes. The

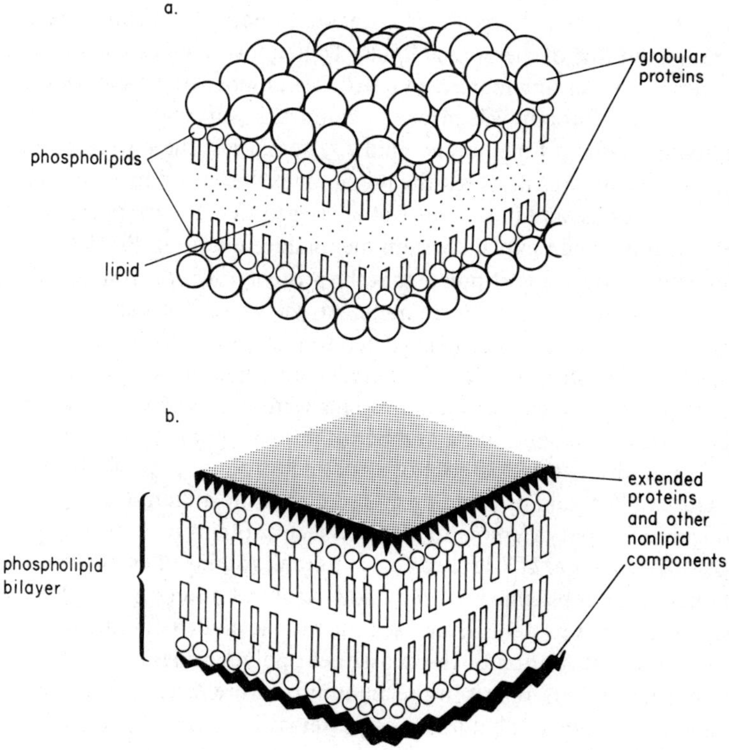

Figure 3.1 Models of the molecular arrangement in cell membranes.
(a) According to Danielli and Davson[7]. The molecules are arranged symmetrically about an undefined lipid layer.
(b) The unit membrane model of Robertson[8]. The arrangement is asymmetric with respect to the nonlipid components coating each surface of a phospholipid bilayer.

essential differences between the unit-membrane hypothesis and the earlier Danielli–Davson model were, firstly, that the secondary structure of proteins coating the external surfaces of the phospholipid bilayer were thought to be in an extended β-structure rather than in globular configuration and, secondly, it was recognised that nonlipid components might be asymmetrically distributed across the membrane. Furthermore, the idea that unspecified lipid material was sandwiched between the phospholipid bilayer was discarded, in order to bring the dimensions of the lipid bilayer into line with more accurate measurements of membrane thickness obtained from X-ray diffraction and electron microscopy.

Although the unit-membrane model was widely acclaimed for several years as a viable model of membrane structure, certain inconsistencies became apparent when membranes other than myelin were examined, particularly membranes containing a high proportion of protein. In this context, we have already seen that most membranes contain insufficient lipid to form a continuous bilayer over the entire surface of the membrane. The unit-membrane hypothesis also states that protein interacts with the hydrophilic groups of lipids predominantly, or even exclusively, through polar forces, and that extensive hydrophobic interactions between the two are excluded. This is clearly incompatible with the necessity for rather drastic procedures to disrupt membranes and with the fact that hydrophobic interactions must be weakened before all membrane proteins can be extracted free of membrane lipids. To help overcome some of the patent deficiencies of the unit-membrane hypothesis a number of different models were advanced in which protein was placed in contact, in varying degrees, with hydrophobic residues of membrane lipids. A number of these were in the form of particulate structures in which the membrane consists of an array of specific lipoprotein subunits, and in some it was envisaged that a rearrangement of the subunit assembly could explain certain membrane functions. Protagonists of lipoprotein models usually stipulate that the protein itself contributes substantially to the overall structure of the membrane. At one time, it was suggested that a separate fraction of specific proteins acted to maintain membrane structure while the remainder were enzymes chiefly concerned with membrane-associated reactions. These so-called structural proteins were supposedly isolated from mitochondrial membranes, but a number of critical appraisals since then have failed to substantiate the existence of structural protein. It would be expected that a structural protein or proteins should constitute an important component of all membranes, but attempts to identify such proteins have generally been unsuccessful. Dreyer *et al.*[9] looked for common proteins in erythrocyte, mitochondrial, sarcoplasmic reticular and retinal-rod outer-segment membranes by extracting all the proteins from each membrane and examining the relative molecular mass distribution of the constituent polypeptides by gel electrophoresis. Despite substantial differences in the amount of protein relative to other membrane components, the number of proteins and their relative molecular-mass distribution was unique to each particular membrane. Erythrocyte and mitochondrial membranes consisted of many proteins, none of which was common to either membrane, while membranes of the sarcoplasmic reticulum and retinal-rod outer segment had only one major polypeptide each. These were identified as Mg^{2+} dependent–Ca^{2+} activated ATPase and rhodopsin, respectively, proteins that are primarily concerned with membrane function and not known to be required for maintenance of membrane structure. It would appear, on these grounds at least, that there is no convincing evidence for the existence of structural protein unless it is conceded that different proteins can act as structural elements in different membranes and, perhaps even more unlikely, that strictly functional proteins play a major role in membrane structure.

As a footnote to the structural protein story, it has been shown that protein from mitochondria, originally believed to be structural protein, proved to be largely denatured protein formed as a result of the particular method used to extract it from the membrane. Another aspect, and one we shall consider in greater detail in section 3.2.1, concerns the apparent freedom of certain proteins to diffuse laterally or to rotate within the membrane. It might be expected that if proteins contributed substantially to membrane structure they would be anchored rather rigidly in position by strong short-range interactions with neighbouring components as well as possibly through weaker interactions with more remote membrane components. This means that considerable energy would be required to displace such proteins in a lateral direction. This does not seem likely, at least for proteins extending into the hydrophobic region of the membrane, which are known to aggregate in the membrane under relatively mild conditions of ionic strength and pH.

3.2 The Fluid-mosaic Concept of Membrane Structure

A comparatively recent concept of membrane structure has been developed by Singer and Nicolson[10] to explain the highly mobile character of individual membrane components[11]. The fluid-mosaic model of membrane structure is illustrated in figure 3.2. The matrix of cell membranes according to this theory is a lipid bilayer to which proteins are either absorbed by predominantly polar forces or are interpolated into the bilayer in direct contact with the hydrophobic

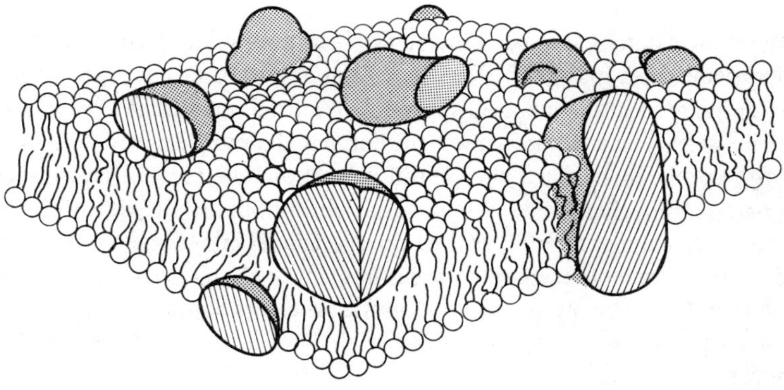

Figure 3.2 The fluid-mosaic model of membrane structure proposed by Singer and Nicolson[10]. Globular proteins are interpolated to varying degrees into a phospholipid bilayer and are distributed randomly in the plane of the membrane.

region of the membrane. This permits short-range ordering, such as that required for the formation of integrated functional units like the multicomponent electron-transport chain of the inner mitochondrial membrane, as well as specific lipid–protein interactions necessary for the activity of many membrane-bound enzymes and antigens. The model presupposes that there is no intrinsic long-range order in membranes and that whenever this can be demonstrated it arises from conditions external to the membrane or from an additive effect of short-range interactions forming an unusually extensive array. The structure is thought to be fluid in the sense that individual molecules are able to diffuse readily in the plane of the membrane to form regions of distinctive affinity or character over the membrane surface. Since mobility is the key feature of this model, it will be worth while to review some of the recent experimental evidence bearing on this point.

3.2.1 Lateral Mobility of Membrane Lipids

We saw in section 2.2 that phospholipids diffuse rapidly in the plane of artificial bilayer dispersions; these studies have since been extended to membranes of biological origin. The lateral diffusion rates of a spin-label phosphatidylcholine analogue incorporated into sarcoplasmic reticular membranes, and dispersions of lipids extracted from these membranes, have been calculated by Scandella *et al.*[12]. The spin-label method relies on the fact that electron spin resonance lines become broadened in a concentration-dependent manner owing to spin–spin interactions between paramagnetic probe molecules. It was found that spin–exchange interactions, resulting from a collision between different probe molecules, was the predominant cause of resonance-line broadening provided that the temperature was higher than about 40°C. The collison frequency Z between spin-label molecules was calculated from spectral-line shapes at various probe concentrations (c) assuming an average area (a) for each lipid molecule in the bilayer of 0.6 nm^2 to give an estimate of the diffusion constant D from the relationship

$$D = \frac{Za^2}{12c}$$

The lateral diffusion of spin-label molecules incorporated into sarcoplasmic reticular membranes (6.0×10^{-12} m^2s^{-1}) was almost as fast as that of the probe interpolated into aqueous dispersions of a total membrane lipid extract (10^{-11} m^2s^{-1}). The latter measurement compared almost exactly with diffusion rates of the probe in artificial phosphatidylcholine–cholesterol bilayer dispersions. It is apparent from these measurements that phospholipids diffuse rapidly in the plane of biological membranes, as they do in artificial bilayer membranes. It should be emphasised that mobility measurements only refer to the lipid domain accessible to the probe molecules and significant errors could arise if probe molecules were restricted to, say, a fluid, as distinct from a rigid, region of the membrane.

Lateral movement of membrane lipids is also indicated from freeze–fracture patterns of membranes from the micro-organism *Acholeplasma laidlawii* prepared in a similar manner to the pure lipid dispersions discussed previously[13] (see section 2.2.3). The membranes of this micro-organism do not normally contain cholesterol, and the fatty acid composition of membrane phospholipids can be selectively enriched by supplementing the growth medium with the desired fatty acids. Thus if the membrane phospholipids contain predominantly unsaturated fatty acids, and the cells are quenched from above the phase-transition temperature, only smooth membrane fracture faces are observed, with membrane-associated particles arranged in a lace-like manner. Conversely, cells quenched from below the lipid phase-transition temperature show particles that are aggregated on the inner fracture face, with a system of ridges in the particle-free regions corresponding with crevices on the complementary fracture face. Similar results have been reported in the ciliate protozoan, *Tetrahymena*[14]. When this organism is cooled from a growth temperature of 28°C down to 5°C, membrane-associated particles become aggregated. If the organism is reheated quickly to 28°C the smooth areas of the fracture face disappear and the particles are again distributed evenly in the plane of the membrane. This demonstrates that the temperature-induced changes in particle distribution within the membrane are completely reversible.

Rearrangement of lipids and proteins in chloroplast membranes following illumination has been reported by Torres-Pereira *et al.*[15]. They used freeze-fracture techniques in conjunction with hydrophobic spin-labels to examine membrane-associated particle distribution and the fluidity of the hydrocarbon region of the membrane. A marked change in the partition of their spin-probe molecule between the aqueous and hydrocarbon region of chloroplast membranes at a temperature of 18°C indicated that the membrane lipids underwent a phase change at this temperature. Using the relationship between temperature and partition coefficient of the spin-label, they were able to show that probe molecules added to chloroplast suspensions that were adapted to dark conditions at constant temperature entered the hydrophobic region of the membrane when they were exposed to light. This change in partition coefficient represented an equivalent increase in temperature of 2°C. They also found that the change in physical state of the membrane lipids was accompanied by an alteration in the distribution of membrane-associated particles in freeze-cleaved preparations. Thus examination of membranes fractured before and after illumination showed that the randomly distributed membrane-associated particles in darkened chloroplasts became aggregated after exposure to light.

The non-random distribution of membrane-associated particles in micro-organisms quenched from below the lipid phase-transition temperature and in chloroplasts exposed to light are more easily explained by migration of the membrane lipids rather than a specific migration of proteins. Nevertheless membrane proteins can be induced to aggregate – usually under conditions that are unlikely to affect the mobility of membrane lipids (see section 3.2.3) – and the possibility of an indirect effect of temperature or light on protein migration cannot be entirely excluded.

3.2.2 Rotational Freedom of Membrane Proteins

Some indication of the freedom of proteins to rotate about an axis perpendicular to the plane of the membrane has been obtained from studies of the visual pigment, rhodopsin. This molecule is a glycoprotein (relative molecular mass about 40 000), which is oriented in disk membranes of the retinal-rod outer segment. From X-ray measurements, the molecule is approximately spherical in shape and has a diameter of about 4.5 nm. It is known that the retinal rod is strongly dichroic, since light polarised perpendicular to the long axis of the rod is strongly absorbed compared with light polarised parallel to the rod. This indicates that rhodopsin is preferentially oriented parallel to the plane of the disk membranes. By contrast, within the plane of the membrane dichroism is transient and lasts only as long as the molecule takes to rotate about its axis perpendicular to the membrane. If rotation is prevented by fixing the molecule in the membrane with bifunctional cross-linking reagents like glutardialdehyde, the chromophore exhibits strong dichroism[16]. In experiments with frog retina, Cone[17] produced intermediate compounds in the bleaching sequence of the rhodopsin chromophore, by exposing the retina to extremely short (5 ns) flashes of light and then measured the decay of the ensuing dichroism with perpendicular and parallel polarised light. By selecting appropriate wavelengths, to distinguish bleaching products from residual unchanged rhodopsin, he was able to measure the rotational relaxation time of the rhodopsin molecule in the membrane. The value he obtained was in the order of 20 ns. From a knowledge of the rate of rotation of rhodopsin the viscosity of the surrounding matrix could be calculated and this was in reasonable agreement with values obtained for the hydrocarbon region of fluid membranes. Tryptophan fluorescence has been explored as an intrinsic probe for measuring rotational kinetics of other membrane proteins and external probes such as eosin have been used successfully in model systems to study the rotation of proteins that have no natural chromophores. In general, these studies indicate that rotational relaxation times for different membrane proteins vary considerably and probably reflect differences in the molecular environment surrounding each particular protein.

3.2.3 Protein Aggregation in Membranes

Changes in the distribution of protein in the plane of the membrane were first noted by Pinto da Silva[18], when he examined freeze-cleaved erythrocyte membrane preparations that had been quenched from acidic media. Membrane-associated particles are usually distributed evenly on complementary inner fracture planes of these membranes at physiological pH but they aggregate when the pH is lowered to 5.5 or less. Adjusting the pH again to 7.4 restores the original particle distribution, so that the process is freely reversible. Changes in particle distribution in the hydrophobic region of membranes appears to be controlled largely by interactions taking place in the aqueous phase, since increasing the electrolyte concentration prevents particle aggregation at low pH. The reason why particles aggregate at low pH is not yet clear but it has been suggested that the situation may be analogous to the precipitation of soluble proteins near their isoelectric point. Apart from high salt concentrations, which interfere with electrostatic interactions between membrane proteins, prior fixation of polar residues with cross-linking reagents like glutardialdehyde also prevents particle aggregation. Cross-linking reactions are believed to anchor proteins more securely in the membrane and prevent their lateral movement in the plane of the membrane. It should be pointed out that treatment of chloroplast membranes with cross-linking reagents does not prevent light-induced aggregation of membrane-associated particles despite the fact that protons are generated during photophosphorylation and that a proton gradient develops across the membrane[15]. This anomaly may be reconciled if it is assumed that protein–protein interactions are responsible for aggregation in the erythrocyte membrane whereas segregation of membrane lipids provides the driving force for the aggregation of proteins in chloroplast membranes.

The aggregation of membrane-associated particles in erythrocyte membranes at low pH also appears to be correlated with changes in the disposition of specific protein receptors exposed on the outer surface of the membrane. Pinto da Silva *et al.*[19] and Nicolson and Painter[20] have used colloidal iron hydroxide and ferritin derivatives of antibodies directed against specific membrane proteins as electron-microscope markers to observe changes in the distribution of surface-receptor sites on the erythrocyte membrane. Colloidal iron hydroxide is known to bind preferentially to N-acetylneuraminic acid residues, most of which are attached to the sialoglycoprotein (see table 1.2) of the red-cell membrane, and the marker was found to be reversibly aggregated at low pH in the same way as membrane-associated particles of the inner-membrane fracture face. Moreover, in other studies reported by Nicolson and Painter[20], antibody prepared against spectrin, an extrinsic protein attached to the cytoplasmic surface of the erythrocyte membrane, not only aggregates this protein on the cytoplasmic surface of the membrane but also causes an aggregation of colloidal iron hydroxide bound simultaneously to the sialoglycoprotein on the outer surface of the membrane. The nature of the coupling between spectrin and sialoglycoprotein is not known but, since spectrin is an extrinsic membrane protein presumably adsorbed to the cytoplasmic membrane surface mainly by electrostatic interactions, the connection between the two proteins is likely to be fairly non-specific. In any event, the results of these experiments suggest a possible mechanism whereby the distribution of cell-surface components can be altered by intra-cellular interactions. The significance of this fact is discussed in the next section.

Another situation believed to involve an interaction between proteins on opposite sides of the membrane is the maturation and budding of enveloped virus from host cells[21]. The viral envelope, although derived from the plasma membrane of the host, does not cross-react immunologically nor does it include any proteins normally associated with the host cell membrane. The mechanism that sequesters viral-envelope proteins into areas of membrane destined to form the viral envelope is thought to be an alignment of another, viral-specific, protein on the cytoplasmic surface of the host membrane. In some instances there is apparently no direct interaction between viral glycoproteins and proteins on the inner surface of the membrane, suggesting that membrane lipids may bridge these two proteins.

The aggregation of a number of cell-surface components has been adduced from studies using specific polyvalent proteins and antibodies directed against cell-surface components. Plant lectins, such as phytohaemagglutinin and concanavalin A, bind specifically to carbohydrate residues of membrane glycoproteins of various cells, including those on the surface of erythrocyte membranes, causing the receptor sites to aggregate. Aggregation of lectin binding sites on the erythrocyte surface is associated with an aggregation of membrane-associated particles in the hydrophobic region of the membrane, but it is not known whether this connection extends to spectrin or other proteins located on the cytoplasmic surface. A number of important cellular reactions such as cell contact and agglutination are believed to depend on the distribution of glycoprotein receptors on the cell surface. Many tissue-culture cells, for instance, stop growing when they reach confluency and will agglutinate and stop dividing when treated with low concentrations of lectin such as concanavalin A. Nevertheless, certain variant cell lines that do not agglutinate and grow normally in the presence of relatively high concentrations of concanavalin A have been isolated[22]. The variant strains usually display the same number of concanavalin A binding sites to which lectin appears to bind with equal affinity but, unlike the parent strains, membrane-associated particles of the inner-membrane fracture face do not aggregate, suggesting that particle aggregation or lectin binding-site distribution is an important factor in cell agglutinability.

An interesting effect on the distribution of lectin binding sites is observed when the cell surface is digested briefly with trypsin. Tryptic digestion causes normal cells to agglutinate and relieves density-dependent inhibition of growth in confluent monolayers of cells in tissue culture. Similarly, mild proteolysis also agglutinates cells from variant lines that are not normally sensitive to lectin and, in the presence of lectin, membrane-associated particlcs of these cells become aggregated. The action of trypsin on normal culture cells has been found to depend to some extent on temperature. Rosenblith *et al.*[23], working with mouse fibroblasts, reported that concanavalin A binding sites were randomly distributed over the cell surface at 4°C or 37°C. These sites remain dispersed after trypsinisation if the cells are kept cold but aggregate if the temperature is raised to 37°C. This implies that fluidity of membrane lipids is necessary for lateral movement of proteins in the membrane and furthermore that removal of certain proteins from the membrane surface permits a multipoint attachment of the binding sites to lectin. Studies by Gunther *et al.*[24] have revealed that lectin-binding sites aggregate because of the multiple binding sites on the agglutinins. By using the agglutination of cells as a measure of the biological activity of concanavalin A, they showed that the native protein with four binding sites is many times more potent than a derivative of the lectin with two binding sites, even though the derivative appears to bind with equal affinity to the same membrane receptor sites. Other effects of concanavalin A, such as its ability to prevent the movement of immunoglobulin receptor sites over the lymphocyte surface, are also lost when the number of binding sites is reduced, suggesting that the multipoint attachment of lectin to a group of binding sites is the primary action of those proteins.

3.2.4 Selective Migration of Membrane Proteins

So far, we have seen that membrane proteins can move in a restricted way in the plane of the membrane. The migration of certain proteins over relatively large distances on the cell surface has, however, been strikingly demonstrated by Frye and Edidin[25] in experiments involving the fusion of cells from established tissue culture lines. They took mouse and human cells, both of which possess immunologically distinct surface antigens (histocompatibility antigens), and fused them with the aid of virus to form a heterokaryon. By visualising the respective mouse and human antigens with an indirect fluorescent antibody technique they were able to observe changes in the location of the antigens contributed by each of the parent cells. Within five minutes of fusion the heterokaryon bore two separate populations of surface antigens, which became progressively mixed until, after about forty minutes, a complete mosaic covered almost the entire surface of the cell. Intermixing of cell-surface components apparently did not require metabolic energy or active protein synthesis, but it did depend on temperature and ultimately on the physical structure of the membrane lipid phase.

The movement of selected surface-receptor sites around the periphery of a number of cell types, including lymphocytes, has now been reported. Thus, in unperturbed lymphocytes, antibody located on the surface of the plasma membrane is distributed at random over the cell surface. If multivalent antibodies directed against these surface immunoglobulins are added to a suspension of cells, the cross-linked antigen–antibody complex migrates to one pole of the cell where it accumulates in the form of a cap[26]. The capping phenomenon can be observed, either directly if fluorescent anti-immunoglobulin is used, or indirectly under the electron microscope with ferritin-conjugated anti-immunoglobulin[27]. The ferritin-conjugated antibody method has shown that capping is preceded by the formation of dense aggregates or patches of antibody bound to immunoglobulin receptors on the membrane surface. It has been suggested by de Petris and Raff[28] that cap formation in lymphocytes results from membrane flow analogous to the mechanism of cellular movement and, to support their contention, they cite the fact that cap formation is accompanied by morphological changes in the lymphocyte such as elongation and uropod formation, features which are usually associated with motile cells. At temperatures above 20°C, immunoglobulin molecules are distributed evenly over the lymphocyte surface and are thought to be maintained as independent antibody receptor sites by thermal agitation. Multivalent antibodies, or lectins (in the case of glycoproteins), effectively cross-link the surface receptors when they come into close proximity and the complexed units then associate into patches located at several points on the cell surface. These events do not require metabolic energy, but are restricted at low temperature when random collisions between binding sites are presumably less frequent. Thermal agitation alone, however, cannot explain the direction of patches into a polar cap because the process is extremely rapid, being completed in less than two minutes at 37°C. Moreover, directional movement of patches to a particular region of the plasma membrane, and the metabolic dependence of cap formation, favour a more specific mechanism. A plausible theory to account for cap formation is to consider that patches constitute regions of restricted fluidity in the membrane and may become anchored to contractile elements in the cytoplasm. Stabilisation of the patch permits other, more fluid, membrane components to flow directionally away from the region in which the cap is formed.

The fate of selectively capped antigens appears to differ depending on the origin of the particular lymphocyte. In some cases the antigen–antibody complex is taken into the cell by endocytosis and can be observed on the inner surface of intracellular vesicles but in other lymphocytes the capped complexes are rarely internalised. This phenomenon has been investigated by Stackpole *et al.*[29], who compared the migration of selectively capped antigens on mouse lymphocytes derived from either the thymus or spleen. Thymus-derived lymphocytes or T-cells are characterised by the presence of a specific antigen, TL, when obtained from the thymus, and can be distinguished from B-cell lymphocytes, when recovered as a mixed T and B cell population from the spleen, by the presence of another surface antigen, θ. B-cells on the other hand are the only lymphocytes from spleen to possess immunoglobulin receptor sites. Each type of lymphocyte participates in different immune reactions (see section 5.1.8) but both have the same histocompatibility antigens (H-2) and lectin binding sites. The capping sites of these antigens and receptors when the cells are exposed to appropriate antibodies are presented in table 3.2. It can be seen that immunoglobulin, θ antigen and concanavalin A binding sites are capped over the Golgi pole of spleen lymphocytes (a mixed population of T and B cells) and they are almost completely removed from the cell surface by endocytosis. In contrast θ, TL and histocompatibility antigens of thymocytes (T cells only) are capped predominantly at a pole opposite to that containing the Golgi apparatus and are only occasionally observed inside the cell. Histocompatibility antigen of spleen cells appears to be exceptional in that capping can occur at either pole and there is some internalisation of the antigen–antibody complex. The overall conclusion from these results is that the pole of the cell to which patched antigen-antibody complexes are directed depends on the type or origin of the lymphocyte rather than on the particular antigen involved. In other experiments, Stackpole *et al.*[29] were able to show that metabolic energy is required for capping irrespective of where the cap is located. Furthermore, cytochalasin B, an agent that disrupts cytoplasmic microfilaments, prevents cap formation of antigens directed towards the pole overlying the Golgi complex but has no effect on capping in thymocytes, where the antigen–antibody complex migrates to the opposite pole of the cell. Clearly, the role of microfilaments in directing cap formation and the significance of the site at which the cap forms in particular lymphocytes remains to be determined.

A selection process also appears to operate when areas of the plasma membrane become internalised during endocytosis or phagocytosis. When alveolar macrophages or polymorphonuclear leucocytes are allowed to ingest inert latex particles or oil droplets, more than half of the original plasma membrane may be taken up into the cell. Analyses of the remaining plasma membrane have shown that certain components are left on the cell surface while others are removed roughly in proportion to the amount of membrane internalised. Tsan and Berlin[30], for example, measured the activities of transport systems for lysine, adenosine and adenine in cells before and after phagocytosis of latex particles and found no decrease in the number of transport sites remaining in the surface membrane. The possibility that new transport sites had been inserted

Table 3.2 The site of cap formation and extent of endocytosis of surface receptors on mouse lymphocytes obtained from the spleen or thymus

Lymphocyte	Receptor	Number of caps			Endocytosis
		Over Golgi	Opposite Golgi	Intermediate	
spleen (mixed T and B cells)	immunoglobulin	28	0	2	++++
	concanavalin A	10	2	0	++++
	θ alloantigen	12	2	0	+++
	H-2	10	11	8	++
thymus (T cells)	TL alloantigen	2	18	5	–
	θ alloantigen	1	9	2	±
	H-2	3	12	5	±

(data from reference 29)

during phagocytosis was discounted in separate experiments in which the transport sites were blocked with non-competitive inhibitors before phagocytosis: the transport processes were apparently still blocked after latex particles had been ingested. Though it must be conceded that the transport sites might be localised in certain areas of the plasma membrane not taken up during phagocytosis, it has been suggested that a selection process operates to segregate those membrane components required for sustaining membrane function from those that are not. A selection mechanism based on function may provide a rational explanation from the standpoint of cell survival, but it is difficult to account for the apparent selection of other components whose function is more obscure. Polymorphonuclear leucocytes, for instance, possess at least three types of lectin binding site, one which binds concanavalin A, another binding a different lectin, *Ricinus communis* agglutinin, and a third site which binds both lectins in a linked manner. Oliver *et al.*[31] found that during phagocytosis there is a loss of total specific lectin binding to the residual plasma membrane which closely parallels the amount of membrane incorporated into the cell. However, only the dual lectin binding site appears to be removed from the cell surface.

There is some indication that microtubular structures in the cytoplasm are involved in some way in the selection of plasma membrane components internalised during phagocytosis. Ukena and Berlin[32] have shown that pretreatment of phagocytes with drugs like colchicine and vinblastine, which disrupt microtubular structures, prevents discrimination of various transport sites internalised during phagocytosis. Likewise, treatment of phagocytes with microtubular-active drugs leads to an indiscriminate uptake of various lectin binding sites during phagocytosis. Since the amount of membrane internalised is unaffected, however, it would seem that intact microtubules are not required for phagocytosis itself. The available evidence strongly suggests that the directional movement of selected components of the plasma membrane is influenced by an interaction with microtubular and microfilamentous structures located in the cytoplasm close to the plasma membrane[33,34].

The fluid-mosaic model of membrane structure is compatible with, and indeed predicts, many of the experimental findings we have considered. The movement of lipid and protein components in the plane of the membrane is now firmly established and, in some instances, appears to be required in certain cellular responses. The movement of membrane molecules over relatively short distances of the membrane surface can be satisfactorily explained by lipid mesomorphism or random thermally induced collisions and affinities between molecules, but the directional movement of other components, often over large distances, is likely to be controlled by factors external to the membrane.

3.2.5 Bifunctional Reagents as Probes of Protein–Protein Interactions

Many multienzyme systems, and particularly those located in the inner mitochondrial membrane, are oligomers of two or more different peptide subunits and the interaction between subunits is presumably required for their functional activity. Not much is known, however, about interactions between other membrane proteins: whether they exist as free and independent polypeptides, interact with each other randomly or are associated in a more specific manner. The distribution of certain antigenic and lectin binding sites on the cell surface and the pattern of membrane-associated particles in freeze-cleaved preparations indicate that there are no strong cohesive forces between different membrane proteins. Nevertheless, interactions between different proteins in the plane of the membrane, as well as between proteins orientated on either side of the membrane, are probably concerned in specific protein migration.

One method of investigating protein–protein interactions in membranes is to cross-link them covalently *in situ* with bifunctional reagents and then examine the type and extent of cross-linking between specific membrane proteins. The procedure can be greatly simplified when cross-linking reactions are reversible because the coupled peptides can be separated and subsequently identified more confidently by gel-electrophoretic mobility. Most of the cross-linking reagents react primarily with amino groups so that proteins can cross-link with amino phospholipids and this may affect the mobility of the peptides in electrophoresis. It should also be pointed out that each cross-linking reagent requires a specific alignment between reactive groups on different peptides, and it is conceivable that even oligomeric subunits may fail to interact if no suitable alignments exist even though the peptides may be in close proximity. Such risks can usually be minimised by employing a number of different bifunctional reagents in parallel experiments.

Steck[35] was one of the first to investigate associations between different membrane proteins deduced from the controlled use of cross-linking reagents. He treated human erythrocyte membrane preparations with formaldehyde, glutardialdehyde and several oxidising agents, of which O-phenanthroline–$CuSO_4$ was the most satisfactory, and then extracted the proteins for an examination of the products formed. Under mild conditions, in which excessive reaction was prevented, he showed that certain bands in gel electrophoretograms diminished or disappeared altogether and were replaced by new high relative molecular mass bands of cross-linked peptides. His general conclusion was that some major proteins of the erythrocyte membrane interacted specifically with identical or closely related proteins, while other proteins did not appear to interact at all. In similar studies of ox erythrocyte membranes, Capaldi[36] found that more extensive cross-linking with glutardialdehyde coupled almost all membrane proteins. Moreover, these highly cross-linked membranes were resistant to disruption by sonication or detergents. By resorting to more drastic procedures such as boiling the membranes in sodium dodecyl sulphate about 10 per cent of the total membrane protein could be removed, virtually all of which was glycoprotein. This confirmed the previous observation of Steck that sialoglycoprotein of the erythrocyte membrane is not cross-linked with glutardialdehyde or, for that matter, by a number of other bifunctional reagents. This suggests that electrostatic repulsion between the highly charged polar residues of these proteins repels other proteins and does not allow them to approach within range of the bifunctional groups.

A refinement of the cross-linking reaction has been reported recently by Ji[37] who used bifunctional reagents, with varying distances separating the functional groups, from which he obtained an estimate of the proximity of proteins in the membrane. In human erythrocyte membranes, for example, glycoproteins were cross-linked with dimethyl adipimidate dihydrochloride in which the functional groups are separated by a distance of 860 pm but not with dimethyl malonimidate dihydrochloride under the same conditions where the distance is only 490 pm. Examination of the complex, cross-linked with the adipimidate reagent, showed that glycoproteins reacted with each other as well as other membrane proteins. These results indicate that glycoproteins of the erythrocyte membrane do not normally approach within 500 pm of other membrane proteins.

Bifunctional reagents have been used to explore the interaction between certain acidic phospholipids and proteins of the human erythrocyte membrane. Marinetti *et al.*[38] treated these membranes with 1,5-dinitrofluoro-2,4-dinitrobenzene or 4,4′-difluoro-3,3′-dinitrodiphenylsulphone, which react primarily with amino groups, and prepared lipid-soluble extracts of the cross-linked membrane. On hydrolysis these extracts yielded water-soluble derivatives of amino-phospholipids, mostly phosphatidylserine and phosphatidylethanolamine. Since about 20 per cent of these phospholipids are cross-linked to membrane proteins whereas only 5 per cent are cross-linked to each other it appears that most amino-containing phospholipids of the erythrocyte membrane are associated with membrane proteins rather than existing in bilayer form.

3.3 Variation in the Surface Topography of Contiguous Membranes

3.3.1 Interactions Between Membranes

Many physiological events rely on contact between different membranes of the cell, or different regions of the same membrane. These contacts can be stable and constitute distinct morphological features such as myelin and retinal rod outer segment membranes or they may be unstable and lead to fusion of the two membranes. The latter group includes such processes as endocytosis and exocytosis in addition to pathological mechanisms like phagocytosis and cellular cytotoxic reactions. Since many specific membrane contacts are restricted to certain regions of the membrane it is often assumed that these areas become specialised in some way to facilitate interaction. We have already seen that glycoproteins on the surface of the plasma membrane must assemble into patches before cells will agglutinate. Another example is the fusion of spermatozoa with ova where contact between the two gametes is established exclusively through the acrosomal region of the sperm. Accordingly the outer acrosomal membrane must possess sites capable of recognising the ovum in addition to components needed to penetrate it once the outer acrosomal membrane has been shed. The distribution of lectin binding sites may be important in this regard because, in the case of the rat at least, it has been shown that concanavalin A binding sites are restricted to the plasma membrane overlying the acrosome[39]. Concanavalin A binding causes the spermatozoa to agglutinate and prevents fertilisation, which suggests that a redistribution site in the membrane surface is involved[40]. It seems that other surface antigens are also located predominantly in the acrosomal region. Koo *et al.*[41], for example, have reported that the male H–Y antigen of mouse, which is responsible for rejection of male mouse skin grafts by female recipients, is only present on the acrosome of mouse spermatozoa and no antigen can be detected on regions of the plasma membrane covering other parts of the cell. The reason for a concentration of these antigens in the acrosomal region is not known but the possibility cannot be entirely excluded that such antigens may prevent multiple sperm penetration of the ovum.

Lectin binding sites seem to be concentrated in certain regions of other membranes. This phenomenon has been demonstrated recently in nerve tissue, where receptor sites for concanavalin A and ricin are located predominantly in synapses[42]. Furthermore, these sites appear to be anchored firmly in the membrane and are not free to diffuse laterally in the plane of the membrane. It is not clear whether such lectin binding sites are involved in membrane interactions but it is known that the membranes of nerve endings in particular are rapidly and continually undergoing fusion.

3.3.2 Intercellular Junctions and Communication

Contact between cells is usually established through specialised junctions, which are considered to be differentiated forms of the plasma membrane rather than seperate organelles[43,44]. Some types of junction involve a high degree of organisation both within the membrane and in the underlying cytoplasm, while others appear to be simply a close apposition of certain regions of the plasma membrane of adjacent cells.

The desmosome (figure 3.3) is a particularly complex form of junction found mainly in epithelial tissues, where it serves to bind neighbouring cells together and provide the tissue with a high degree of mechanical resilience. The plasma membranes of adjoining cells align parallel to each other, and the membranes are separated by an interspace of between 25 and 30 nm. Certain tracer molecules such as colloidal lanthanum hydroxide are able to penetrate between the membranes so the interspace is, to some extent, continuous with the extracellular

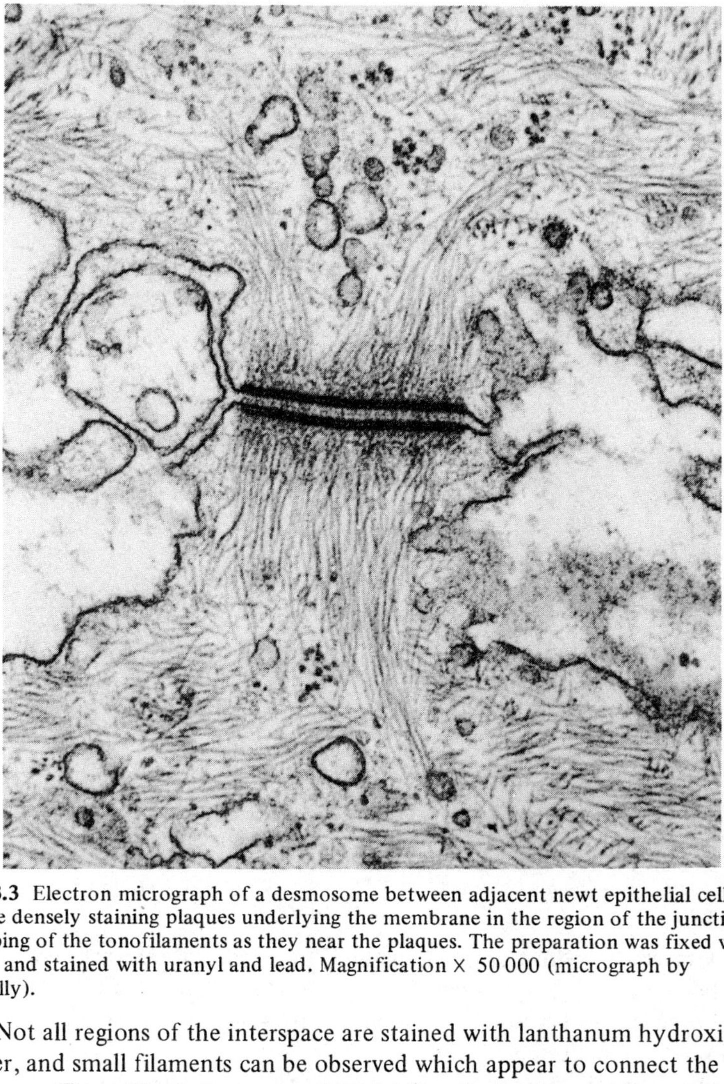

Figure 3.3 Electron micrograph of a desmosome between adjacent newt epithelial cells. Note the densely staining plaques underlying the membrane in the region of the junction and the looping of the tonofilaments as they near the plaques. The preparation was fixed with osmium and stained with uranyl and lead. Magnification × 50 000 (micrograph by D.E. Kelly).

space. Not all regions of the interspace are stained with lanthanum hydroxide, however, and small filaments can be observed which appear to connect the two membranes. These filaments presumably stabilise the junction and may also act as anchoring sites for the cytoplasmic structures, referred to as tonofilaments, which radiate out into the cytoplasm from the region of the desmosome.

Zona occludens, or tight membrane junctions, maintain close contact in a continuous belt-like region at the apex of adjoining columnar cells such as those lining various glands and tubular organs of the body. In some instances the junction is not continuous and may even be restricted to small areas of contact in which case they are referred to as *macula* or *focal occludens.* The *zona occludens* prevents molecules like haemoglobin (relative molecular mass 64 500) or horseradish peroxidase (relative molecular mass 40 000) passing from the lumen into the intercellular space between adjacent cells of the luminal periphery, and the interspace between the membranes of the junction itself is not accessible to colloidal lanthanum hydroxide (see figure 3.4). The intercellular space is, however, accessible to low relative molecular mass solutes, but only if they enter

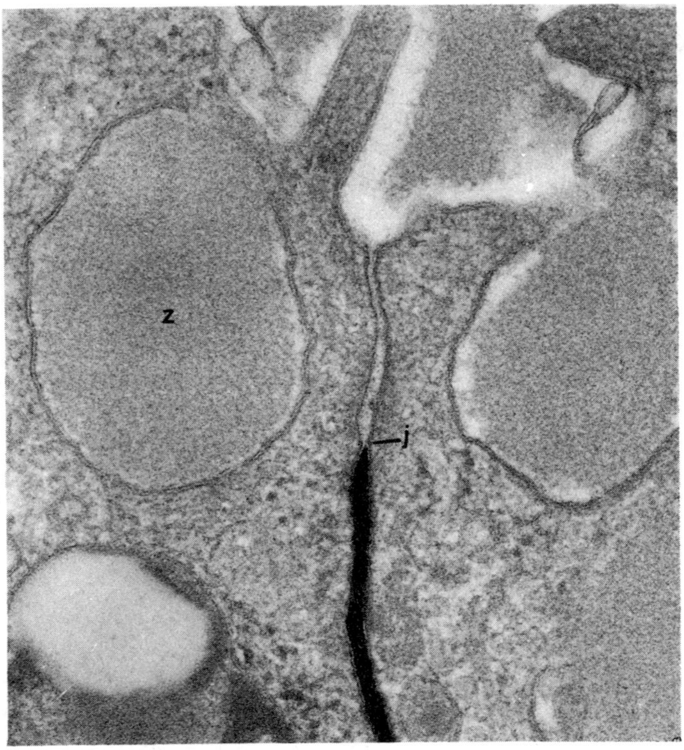

Figure 3.4 Electron micrograph of a thin section through a rat pancreatic acinar cell from tissue injected with lanthanum hydroxide before fixing with paraformaldehyde–glutardialdehyde containing potassium pyroantimonate, sectioning and staining with uranyl acetate and lead acetate. The penetration of lanthanum into the intermembrane space from the vascular pole of the cell is impeded by the distal end of a *zona occludens* junction (j) between the cells. A zymogen granule (Z) is indicated. Magnification × 65 000 (micrograph by D.S. Friend).

via an extracellular pathway, indicating that the intermembrane region is not completely occluded. *Macula occludens* type junctions can be readily distinguished from *zona occludens* because large molecules are able to penetrate into the intercellular space of the former.

The characteristic ultrastructural feature of *zona occludens* is that the total thickness of the junction (14 nm) is usually less than the sum of the thicknesses of each of the two plasma membranes (15 nm), indicating that the membranes are in extremely close contact. Freeze–cleave replicas fractured along the plane of these junctions reveal a network of ridges on one fracture plane and a corresponding system of grooves on the other[45]. In urinary bladder and gall bladder, membrane-associated particles located in the junction form fibrils connecting the outer leaflets of the adjacent membranes. The complex appearance of the *zona occludens* suggests that a considerable rearrangement of membrane components must take place in the formation of this type of junction.

One of the most common forms of cell junction is the gap junction or nexus. These junctions are found in all tissues, including cardiac and smooth muscle, epithelial and connective tissues and at electrical synapses in the nervous system.

Gap junctions provide points of strong adhesion between cells, a feature that has been elegantly demonstrated by Muir[46] in cardiac muscle. When these cells are suspended in a calcium-free medium they separate at all points except gap junctions. The cells will eventually separate completely if deprived of calcium for long periods but the junctions remain intact on one cell leaving a corresponding hole in the membrane of the previously adjoining cell. Gap junctions are thought to provide a passage for communication between cells because, in experiments where microelectrodes have been inserted into the cytoplasm of connecting cells, the junction appears to form a pathway of low electrical resistance between them. Conversely, if one electrode is placed in the extracellular space, a characteristically high electrical resistance is recorded, suggesting that the permeability of the plasma membrane in the region of the junction is different from other areas of the membrane. It could be argued that electrical coupling between cells such as cardiac and smooth muscle are required to elicit a synchronous response in these tissues when subjected to appropriate stimuli. In other cells, however, such as hepatocytes, which do not propagate action potentials, gap junctions probably serve a different function. It has been shown, for example, that gap junctions permit the transfer of certain metabolites from the cytoplasm of one cell to that of another.

The passage of molecules between cells is bidirectional and it has been calculated that the flux of nucleotides across gap junctions occurs at a rate of between 10^5 and 10^6 molecules per second, which, by way of comparison, is enough to construct a complete chromosome of *Escherichia coli*. The passage or exclusion of small dye molecules has been used to estimate the diameter of the channels or pores and an upper limit of 1 nm has been suggested[47]. A large number of other compounds are known to pass across these junctions, but in general only molecules of relative molecular mass less than about 1000 can be transferred efficiently. Metabolic coupling via gap junctions has been reported by Gilula *et al.*[48] in studies of hamster fibroblasts in tissue culture. Normal fibroblasts are able to incorporate hypoxanthine into DNA with the aid of the enzyme inosinic pyrophosphorylase. When these cells are grown in mixed culture with a variant fibroblast line that is unable to utilise hypoxanthine and lacks inosinic pyrophosphorylase, the variant cells can incorporate hypoxanthine into DNA provided they are connected to normal fibroblasts by gap junctions. Variant fibroblasts that are unable to form gap junctions with normal cells do not incorporate exogenous hypoxanthine. Apart from rectifying metabolic deficiencies, gap junctions may also be concerned with other forms of metabolic co-operation between cells such as inductive interactions during tissue differentiation and embryogenesis.

An interesting series of experiments described by Azarnia *et al.*[49] suggests that the ability of cells to form gap junctions is genetically controlled. They fused a human cell line, in which gap junctions are the only discernible form of cell contact, with a mouse cell line incapable of establishing any kind of cell junction. All the hybrid cells formed electrically coupled gap junctions and they retained this ability provided they had an almost complete set of parent chromosomes. Some clones derived from among the segregants again lost the ability to form gap junctions and these revertants were invariably deficient in human chromosomes.

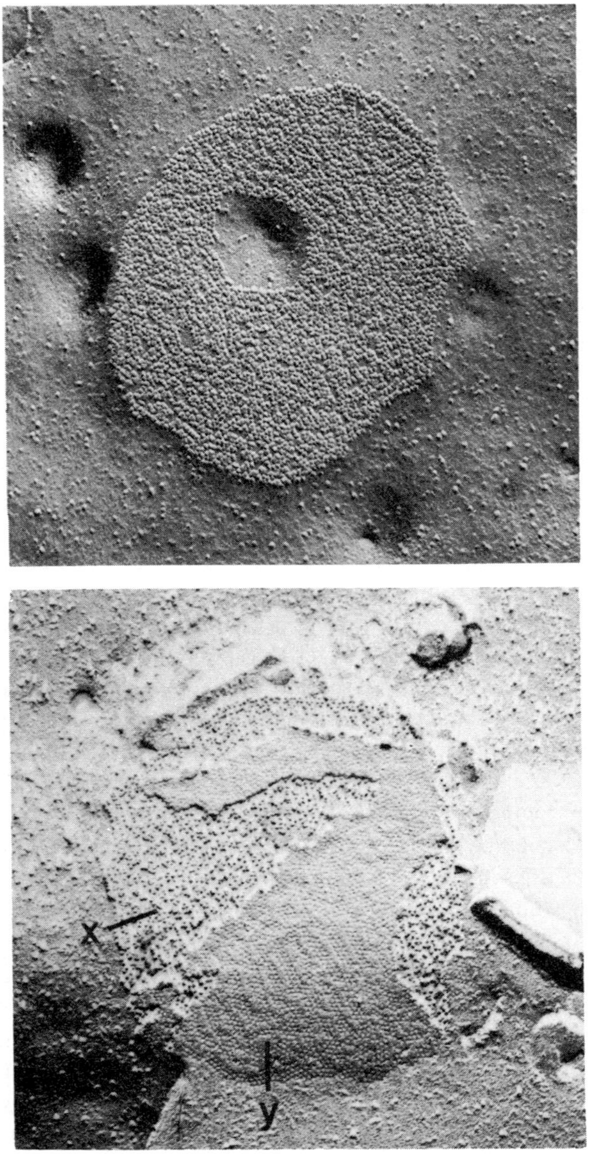

Figure 3.5 Freeze-cleaved preparations of gap junctions between cells. The first preparation (a) shows a single fracture plane along a gap junction connecting two 3T3 mouse fibroblasts. Note the high density of particles in the region of the junction. The replica of the second preparation (b) shows a gap junction between two SV40 virally transformed 3T3 mouse fibroblasts. In this case two membrane planes are exposed, one containing particles protruding above the surface (x) and the other with pits (y). Magnification × 50 000 (micrographs by S. Knutton).

It was concluded that human chromosomes supply a genetic factor which corrected a deficiency in the mouse cell line, namely factors responsible for gap-junction formation, in a manner suggesting that the mouse cells were genetically recessive for this character. How genetic control of membrane interactions is exercised is not known but synthesis of specific membrane proteins concerned in the junctional complex must be considered a likely possibility.

The ultrastructure of gap junctions in cardiac muscle cells has been examined by McNutt and Weinstein[50] using thin section and freeze–cleave electron microscopy. In thin section, the junction appears as a parallel alignment of plasma membranes separated by a space of 2 to 3 nm. This space is penetrated by colloidal lanthanum hydroxide to reveal a complex subunit structure. In transverse section, globular subunits about 7 to 8 nm in diameter are separated by a centre-to-centre spacing of 9 to 10 nm and, when seen *en face*, the subunits are arranged in a characteristic hexagonal array. Larger subunits (8.5 to 10 nm diameter) are found in gap junctions connecting invertebrate cells and occasionally in higher organisms but, as in the more common form, they are always arrayed hexagonally in the plane of the junction. Freeze–cleaved preparations show a closely packed array of particles 5 to 7.5 nm in diameter protruding about 5 nm above the plane of the cytoplasmic membrane leaflet (figure 3.5). Small depressions are sometimes observed in the centre of these particles and these are thought to represent the site of channels extending through the membrane.

In contrast to plasma membranes in general, gap junctions are surprisingly resistant to disruption with certain detergents. This indicates that the type or particular arrangement of membrane components in the region of the junction may be different. From an operational viewpoint this property has provided a convenient method whereby gap junctions can be isolated and analysed. Mouse-liver gap junctions isolated by detergent treatment appear to have a higher proportion of neutral lipid and a distinctly different phospholipid composition from plasma membrane[51,52]. There is generally more lipid relative to protein and a single peptide of relative molecular mass about 20000 predominates. Some restraint is warranted in accepting such results as being other than preliminary analyses, especially in view of the fact that detergents may selectively remove some membrane components from the junction complex without necessarily altering the morphology of the membranes. Nevertheless, if there are any true differences in membrane composition this must to some extent reflect the need to preserve the integrity of membranes that are in close contact and, while comparisons between gap junctions and myelin may not be strictly correct, it is noteworthy that the composition of this membrane differs markedly from the plasma membranes of other cells. By analogy, one might also expect the composition of the membranes in regions involved in unstable interactions and fusion to be altered, but there is as yet no evidence available to substantiate this view.

3.4 Structural Asymmetry in Membranes

Early models of membrane structure depicted a symmetrical arrangement of proteins coating either side of a phospholipid bilayer (see figure 3.1a). With the advent of the electron microscope, detailed studies of membrane ultrastructure indicated that many cell membranes were unlikely to be completely symmetrical; the unit-membrane hypothesis was and attempt to incorporate this feature into a model of membrane structure. Myelin, for example, consists of uneven layers of membranes corresponding to the alternate apposition of cytoplasmic and external surfaces of the Schwann cell plasma membrane. (see figure 1.2). Because the spaces between the closely aligned membranes expand to different extents in media of low ionic strength it is generally believed that different chemical groups are exposed on either side of the membrane. Other examples of membrane asymmetry have been observed in preparations stained with osmium or permanganate, which occasionally show distinct morphological features associated with one or other surface of the membrane; particles attached to the matrix surface of the inner mitochondrial membrane provide a good example. Occasionally layers are present that coat either or both surfaces of membranes. Such material on the cell surface does not seem to be essential for survival because it can often be removed by gently washing cells in moderately high or low salt concentrations without serious detriment to the cell. Although these layers are rapidly restored on the cell surface it is doubtful whether they can strictly be regarded as integral components of the membrane.

Freeze–cleave membrane preparations also show asymmetry with respect to the density of membrane-associated particles on complementary fracture planes. Although the nature of the intermolecular forces which determine the distribution of particles between the two membrane leaflets is not clear, the fact that asymmetry is observed implies that these forces are not the same on both sides of the membrane. Furthermore, the particle density of particular membrane fracture planes appears to be relatively constant, so that the distribution is a permanent rather than a transient membrane feature. The particles themselves are probably interpolated proteins, but it is not certain that they represent intact proteins, because there is some evidence that covalent bonds can be broken by the shearing forces generated during membrane cleavage.

More direct evidence of the asymmetric distribution of membrane components has been obtained by observing the sites to which specific antibodies and lectins bind to the membranes. These agents can be conjugated to certain markers such as ferritin without loss of binding affinity and are easily identified under the electron microscope[53]. The mitochondrial membrane is another example of structural asymmetry, and a functional sidedness has also been demonstrated in this membrane. We shall have an opportunity later (section 5.3.6) to examine the topographical asymmetry of the inner mitochondrial membrane.

3.4.1 The Cell Coat

The presence of carbohydrate on the surface of most cells has been well documented, mainly on the basis of histochemical evidence[54]. Cells treated with periodic acid–Schiff's reagent, a specific carbohydrate stain, usually show a denser staining layer on the outer surface of the plasma membrane while the cytoplasmic surface of the membrane and most other subcellular membranes do not take up the stain. Variations of this staining method, suitably adapted for electron microscopy, however, have shown the presence of carbohydrate on the inner surface of Golgi membranes and it has been proposed that carbohydrates are attached to peptide precursors at this site in preparation for their ultimate transposition into plasma membrane.

Asymmetry of charge distribution across the plasma membrane has been investigated by the ability of certain cationic markers such as colloidal iron hydroxide, Alican blue and ruthenium red to adsorb to the membrane surface. The markers bind mainly to ionised carboxyl and sulphate groups of acidic carbohydrates, which appear to be located predominantly on the external surface of the membrane. Removal of sialic acid residues from the membrane surface by treatment of cells with neuraminidase markedly reduces the binding affinity for cationic dyes, so that these residues are likely to contribute substantially to the overall surface charge of the membrane. Wallach and Eylar[55] have also noted that treatment of ascites tumour cells with neuraminidase substantially reduces their anodic mobility when placed in an applied potential gradient, which tends to confirm this view. Nevertheless, acidic carbohydrates probably constitute only a small fraction of the total membrane carbohydrate. Some years ago Gasic and Gasic[56] compared the staining intensity of periodic acid–Schiff's reagent with colloidal iron hydroxide binding to the surface of ascites tumour cells. After the cells had been treated with neuraminidase, colloidal iron hydroxide was, as expected, no longer bound to the surface but the periodic acid–Schiff's reaction persisted almost to the same intensity as that in untreated cells. They were able to remove the periodic acid–Schiff's reacting material by treating the cells with glycosidase enzymes, confirming that the stain reacted specifically with neutral carbohydrates. Since treatment of cells with trypsin could remove most of the acidic and neutral sugars from the cell surface, glycoprotein rather than glycolipid appears to be the predominant source of carbohydrate of the cell coat.

3.4.2 Protein Disposition Examined by Controlled Proteolysis

When intact cells are treated with proteases, proteins exposed on the outer surface of the plasma membrane are selectively cleaved[57]. In most cases the digested membrane remains intact and impermeable to the enzyme despite the fact that considerable amounts of peptide may be released. This means that intracellular proteins, including those exposed on the cytoplasmic surface of the membrane, are protected from attack. Unfortunately some cell types do not behave in this way and are rapidly lysed by tryptic digestion. A number of tissue-culture cells and lymphocytes fall into this category and consequently membrane protein disposition cannot be examined by this procedure. After protease treatment, membranes can be isolated and specific peptides identified by comparison with gel-electrophoretic patterns of undigested membranes. Controlled proteolysis can also be applied to membrane preparations in the form of resealed vesicles able to exclude enzymes, but precautions must be taken to ensure that all the membrane is orientated in the same way. Protease digestion of resealed membrane preparations has shown that, in general, relatively more protein can be released from the surface of these vesicles than from an equivalent surface area of plasma membrane of intact cells.

The type of peptides released obviously depends on the particular enzyme used. Many proteases have been used for this purpose including endopeptidases such as *Bacillus subtilis* proteases, *Streptomyces* protease, papain, bromelain and ficin, all with non-specific action, as well as specific endopeptidases like trypsin, which hydrolyses bonds adjacent to arginine or lysine, and chymotrypsin, which attacks peptide linkages involving aromatic amino acids. The removal of terminal amino acid residues with carboxypeptidases A and B and leucine aminopeptidase from membrane proteins has also been investigated. Trypsin, chymotrypsin and papain have all been used by Steck *et al.*[58] to investigate protein asymmetry in erythrocyte membranes. They monitored the rate and extent of digestion of membrane vesicles, prepared with normal and inverted orientations, by measuring the release of protons during the reaction. Judging from the amount and rate of proton generation, more proteins are susceptible to hydrolysis on right-side out vesicles than on inverted vesicles. Furthermore, according to the polyacrylamide gel electrophoretograms, all the major membrane proteins (with one exception), including the sialoglycoproteins, were vulnerable to attack in normally orientated vesicles as well as membrane ghost preparations, whereas only one peptide and the sialoglycoproteins were accessible in inverted vesicles. It was shown that the peptide not degraded in right-side out membrane was also protected from hydrolysis in inverted vesicles, indicating that it was located at an internal membrane site not accessible to proteases from either surface.

Controlled proteolysis of the erythrocyte membrane of the type outlined above clearly shows that this membrane does not consist of a symmetrical arrangement of proteins coating either surface of a phospholipid bilayer. On the contrary it provides evidence that at least some proteins penetrate completely through the bilayer, since they are susceptible to protease attack from both surfaces of the membrane. The method does, however, suffer some serious disadvantages including the possible rearrangement or loss of proteins during membrane isolation or when preparing membrane vesicles with the required orientation. This may well explain why right-side out vesicles release more peptides compared with intact erythrocyte membranes. One method that has been devised to help overcome this difficulty is to form derivatives of membrane proteins in intact cells before digesting the membrane in the usual way. The products released can be identified as particular derivatives of chemically modified proteins and their disposition may be compared with the products released from sealed membrane-vesicle preparations.

3.4.3 Labelling Methods for Assessing Membrane-protein Disposition

During the past few years a number of chemical agents have been developed that, by virtue of their size, charge or other polar feature, are not readily permeable to membranes yet are capable of reacting vigorously and, in many cases, specifically with membrane proteins exposed on the membrane surface. These labels are usually designed so that extremely small amounts can be detected by either radiochemical, fluorescence or electron paramagnetic techniques. Some reagents have the added advantage of being reactive under relatively mild conditions of pH, temperature and ionic strength so that membrane topography is not unduly perturbed and cell lysis is prevented. Most labelling reagents have a finite permeability through membranes and, in so doing, react with proteins on the inner surface of the membrane as well as cytoplasmic proteins. The reaction time must therefore be carefully controlled so as to accentuate the reaction with outer-membrane proteins while at the same time minimising labelling of cytoplasmic proteins. A useful reaction index has been to compare the relative specific labelling of outer surface-membrane proteins with typical cytoplasmic proteins; haemoglobin, for example, has been used as a convenient marker in erythrocyte-labelling experiments.

A farily impermeable radioactive reagent, [^{35}S]-formylmethionylsulphone methyl phosphate, has been prepared by Bretscher[59] and used to investigate the disposition of proteins of the human erythrocyte membrane. The reagent reacts primarily with amino groups, and possibly hydroxyl groups as well. Labelled proteins of intact and ruptured membrane preparations were extracted and resolved by polyacrylamide gel electrophoresis and further characterised by amino acid fingerprinting techniques. A single protein appeared to be labelled from both sides of the membrane, suggesting that it extended completely through the membrane and resided in a fixed orientation with respect to the inner and outer surfaces of the membrane. Another impermeable reagent of this type is the diazonium salt of [^{35}S]-sulphanilic acid, which also reacts with protein amino groups. Unfortunately the usefulness of this reagent is restricted because the resulting protein derivatives cause the membrane to become leaky to sodium and potassium ions and the cells eventually lyse[60].

A different procedure has been introduced by Rifkin *et al.*[61] involving the formation of a Schiff's base between free amino groups of protein and pyridoxyl phosphate followed by reduction with NaB^3H_4. They argued that since pyridoxyl phosphate is impermeable to most membranes the initial pyridoxylation, and consequently the labelling reaction, should be confined exclusively to proteins on the outside of the cell. They tested this on influenza virus, which contains five membrane proteins, four of which are glycoproteins, exposed on the outer surface of the envelope; the remaining protein is believed to be associated exclusively with the inner surface of the membrane. They were able to confirm that only glycoproteins exposed on the outside of the viral envelope were labelled and that all these could be removed by protease digestion of the intact virus particles. Probably the greatest disadvantage of the method is that protein amino groups vary in reactivity to pyridoxylation, so that it is difficult to compare the proportions of different membrane proteins.

A similar method for labelling surface sialic acid residues has been reported by Blumenfeld *et al.*[62]. In this case the sugar is converted to an aldehyde by mild oxidation with periodic acid and then reduced to 3-deoxy-5-acetamidoheptulonic acid with NaB^3H_4 in order to introduce the radioactive label. Membrane proteins of the erythrocyte, other than sialoglycoprotein, were not significantly labelled in intact cells although various lipids including glycolipids, plasmalogens and lipid peroxides were all extensively labelled. It was also reported that labelling efficiency was low in intact cells compared with isolated membrane preparations, suggesting that perturbation of membrane structure may improve the conditions for oxidation–reduction reactions.

As we have already noted, the main objection to the use of low relative molecular mass reagents is that they can and do penetrate the membrane slowly and react with intracellular proteins. Tedious measurements of membrane permeability or comparison of labelling patterns of intact cells with those of lysed preparations do not always provide a satisfactory answer to this problem. One method used to overcome this difficulty is to employ labelling reagents of high relative molecular mass that are unquestionably too large to penetrate the membrane. A popular procedure is to link a low relative molecular mass reagent to a large sugar polymer. The advantage of carbohydrate polymers is that their size can be varied as required and, in reactions where the label becomes covalently attached to the membrane, the sugar residues can subsequently be removed with glycosidases or left *in situ* to serve as specific antigenic sites. Lipid-soluble reagents have also been adapted for use in labelling studies. Schmidt-Ullrich *et al.*[63] describe a fluorometric method for labelling external proteins using dansyl chloride, a reagent commonly used for N-terminal analysis of proteins (see section 1.3.6). The reagent is converted to an unreactive form in free solution and must be protected by placing it in an apolar environment. This was achieved by incorporating dansyl chloride into an aqueous dispersion of dipalmitoylphosphatidylcholine–cholesterol mixtures, which serves as the reagent vector. When the dye-complex was incubated with human or sheep erythrocytes for periods of up to three hours, membrane proteins were substantially labelled but there were no apparent changes in the distribution of peptides on gel electrophoretograms and the physical properties of the membrane, as judged by osmotic fragility, were unchanged. Moreover, haemoglobin did not react in intact cells and only those peptides that label with other reagents were dansylated. Similar results were reported when isolated red-cell membrane preparations were labelled by this method.

An alternative procedure using high relative molecular mass reagents is to employ enzymes to modify sites exposed on the membrane surface and then label these sites in a subsequent step. Gahmberg and Hakomori[64] have developed an enzymic method for labelling membrane glycoproteins based on a variation of the borohydride reduction procedure[61,62]. The primary hydroxyl groups of galactose and N-acetyl galactosamine were oxidised with galactose oxidase (relative molecular mass 75 000) and the products then reduced by NaB^3H_4. Glycoproteins of intact erythrocytes were labelled by this procedure and no cell lysis was detected. Surprisingly, when isolated membrane preparations were reacted, an additional glycoprotein of relative molecular mass 150 000 becomes labelled, suggesting that in intact cells this protein is unreactive because of its location on the cytoplasmic surface of the membrane. A more likely explanation, however, is that the glycoprotein becomes accessible to oxidation as a result of its rearrangement in the membrane during cell lysis.

Surface proteins have also been labelled with radioactive iodine using the enzyme lactoperoxidase (relative molecular mass 78 000), which, in the presence of peroxide, iodinates exposed tyrosyl residues. The advantage of labelling with isotopes of iodine (^{125}I or ^{131}I) is that radioactivity is emitted in the form of γ-rays, which are easy to detect in conventional γ-spectrometers, and the samples can be assayed directly without laborious preparation. Phillips and Morrison[65] first used this technique to label intact erythrocytes and found that only the major sialoglycoprotein and another glycoprotein of relative molecular mass

90 000 become iodinated, in contrast with isolated membrane preparations where most of the proteins become labelled. Reichstein and Blostein[66] adopted a novel approach to the study of protein disposition in the red-cell membrane by using both radioactive isotopes of iodine. The inner surface of the erythrocyte membrane was labelled with iodine by trapping lactoperoxidase inside lysed membrane preparations and then resealing them. After reaction with one iodine isotope the cells were washed to free them of excess free iodine, and then proteins exposed on the outer surface of the washed membranes were labelled with the other isotope of iodine. Comparison of the radioactive profiles of peptides separated by gel electrophoresis confirmed previous observations that a peptide of apparent relative molecular mass 90 000 was labelled from both sides of the membrane.

A consistent feature of lactoperoxidase labelling is that a large proportion of potential labelling sites on the erythrocyte membrane appear to be inaccessible to the enzyme. Proteins extracted from the membrane with butanol, for example, are much more reactive and all fractions become labelled[67]. Moreover, quantitative estimates of the number of iodinated sites in intact membranes indicate that only about 2 per cent of the putative reaction sites become iodinated even assuming that there is only one reactive tyrosyl residue per glycoprotein molecule and no di- or tri-iodotyrosyl derivates are formed. A possible explanation for the inefficient labelling is that the carbohydrate coat on the outer surface of the membrane prevents the enzyme from reaching most of the potential labelling sites. This view is consistent with the finding that removal of 20 per cent of the total sialoglycoprotein from the membrane surface with trypsin results in a tenfold increase in lactoperoxidase labelling; removal of further protein causes a corresponding decrease in iodination. According to the current hypothesis, partial removal of the carbohydrate coat exposes tyrosyl residues on other membrane proteins, but since protease also removes potentially reactive groups the number of labelling sites decreases. If the same assumptions are made as before, the number of potential sites that become labelled even under these conditions is still only about 25 per cent, so many tyrosyl residues must still remain inaccessible to the enzyme. Lactoperoxidase labelling of the plasma membrane of other cell types including lymphocytes, normal and virally transformed fibroblasts in tissue culture and certain viruses have shown that, like the erythrocyte membrane, proteins are arranged asymmetrically in the surface membrane.

The outer membrane of intact mitochondria has also been iodinated with lactoperoxidase[68]. Gel electrophoresis of outer-membrane peptides showed that all twelve fractions were labelled, but a peptide of relative molecular mass 14 000 had by far the highest specific activity, indicating that this membrane also has an asymmetric arrangement of proteins. Iodination of proteins situated in the outer mitochondrial membrane does not appear to alter the permeability characteristics of the membrane because there is no loss of adenylate kinase activity. This enzyme is located in the intermembrane space and its loss from the mitochondria would be one of the first signs of a change in membrane permeability. Furthermore, the exclusion of lactoperoxidase from the intermembrane space is inferred from the fact that specific labelling of the outer membrane is increased twelvefold when it is separated from the inner membrane and matrix proteins. It was found, however, that some labelled protein remains with the inner membrane, suggesting that the digitonin treatment employed to separate the two membranes releases only selected regions of the outer membrane, leaving some parts of it attached to the inner membrane. Moreover, the labelled peptides of the outer membrane removed by digitonin differed from those remaining attached to the inner membrane, and the labelling patterns were both distinct from either the total membrane extract of intact mitochondria or preparations of inner membrane labelled separately with lactoperoxidase. It was suggested that regions of the outer mitochondrial membrane, possibly those in more intimate contact with the inner membrane, are not removed by digitonin treatment. This implies that proteins susceptible to iodination by the lactoperoxidase procedure are not randomly distributed in the plane of the outer mitochondrial membrane.

3.4.4 Asymmetry of Lipid Bilayers

There is mounting evidence that membrane lipids as well as proteins are asymmetrically distributed between the two leaflets of lipid bilayers. Studies of model membrane systems, such as small bilayer vesicles formed from mixtures of different phospholipids, have indicated that the distribution of the constituent phospholipids between the inner and outer bilayer leaflets is not symmetrical. So far, it appears that segregation is most pronounced in small vesicles with low radii of curvature and this is believed to result from the need to pack each molecule into the bilayer in a configuration of lowest free energy. The ability to discriminate phosphorus and proton magnetic resonance signals of particular phospholipids in mixed dispersions, together with the introduction of techniques capable of establishing the location of particular chemical groups on the inner or outer surface of the vesicles, has greatly expanded our knowledge in this area.

Michaelson *et al.*[69] have investigated phospholipid asymmetry in cosonicated bilayer vesicles consisting of equimolar amounts of phosphatidylcholine and phosphatidylglycerol. Both phospholipids possess characteristic phosphorus resonance signals, and a strong methyl proton signal from choline further distinguishes phosphatidylcholine from phosphatidylglycerol, which has unique proton resonances arising from the nonexchangeable protons of carbon-2 and carbon-3 of the terminal glycerol residue. Resonances from nuclei on the outer surface of the bilayer vesicles were distinguished from those located on the inner surface by the addition of impermeant paramagnetic ions that selectively perturb the signals from accessible groups on the outer surface of the bilayer. Paramagnetic ions such as manganese (Mn^{2+}), europium (Eu^{3+}) and praseodymium (Pr^{3+}) modify the resonance signals in a characteristic manner depending on the particular ion, the environment in which they are located and the nature of the chemical groups with which they interact. Michaelson and his colleagues used manganese as the perturbing ion and recorded the effects of increasing manganese concentration on phosphorus and proton magnetic resonances of their cosonicated dispersions as well as separate dispersions of each phospholipid. They found that phosphorus and $-^{+}N(CH_3)_3$ proton resonances of phosphatidylcholine dispersions and the phosphorus resonance of phosphatidylglycerol vesicles were broadened in a biphasic manner with increasing manganese concentration, indicating that the lipid was in the form of bilayer vesicles with approximately 60 per cent of the molecules residing in the outer bilayer leaflet. The phosphorus resonance of phosphatidylglycerol vesicles was found to broaden at relatively low manganese concentrations, which is consistent with the existence of a strong electrostatic attraction between the paramagnetic ion and ionised phosphate groups of the phospholipid. At high manganese concentrations, resonances arising from molecules located on the inner surface also became progressively broadened, suggesting a breakdown of the permeability barrier to the ion. When manganese was added to cosonicated vesicles of the two phospholipids the phosphorus resonance was again reduced by 60 per cent, but in this case the contribution of phosphatidylglycerol to the broadening effect was greater than that of phosphatidylcholine, indicating a predominance of phosphatidylcholine molecules on the inner leaflet. They were able to obtain a more precise estimate of this distribution by observing the —$^+N(CH_3)_3$ proton resonances; only 40 per cent of this signal was affected by external manganese, giving a ratio of phosphatidylglycerol:phosphatidylcholine of 2:1 in the outer leaflet of the bilayer.

Similar experiments have also shown that cholesterol and egg phosphatidylcholine pack asymmetrically into bilayers of cosonicated lipids[70]. In these studies, —$^+N(CH_3)_3$ proton resonances of phosphatidylcholine vesicles containing various proportions of cholesterol were perturbed by adding praseodymium ions to the lipid dispersion. The vesicles are impermeable to praseodymium which, in contrast to manganese, causes a selective shift in the —$^+N(CH_3)_3$ proton resonance signal to a new position of lower field strength rather than producing a decrease in signal amplitude. Measurements of signal amplitude of shifted (outer leaflet) resonances of pure phospholipid vesicles indicated that about 69 per cent of molecules are located in the outer leaflet of the bilayer. This proportion was unchanged by including up to 25 mole per cent of cholesterol in the bilayers, but when the amount of cholesterol was increased to 33 mole per cent there was a marked decrease in the magnitude of the chemical shift produced by praseodymium. This different chemical shift was interpreted as a change in the environment surrounding the choline residues when phospholipid molecules are 'condensed' by cholesterol. With vesicles of high cholesterol:phospholipid ratios there was also a pronounced change in the relative signal amplitudes of shifted and unshifted resonances equivalent to a transfer of 5 per cent of phospholipid from the inner to the outer bilayer leaflet. When these experiments were repeated with dipalmitoylphosphatidylcholine an apparent redistribution of phospholipid molecules was not observed even when the proportion of cholesterol was increased to 40 mole per cent. Nevertheless the change in chemical shift, thought to be associated with the condensation effect, was still manifest, but the transition was observed with a much lower proportion of cholesterol.

The asymmetry of acidic phospholipids or cholesterol across mixed bilayer preparations in the experiments outlined above could be explained on the basis of the probable constraints imposed on molecules forming planar structures of low radii of curvature. In the case of mixed phospholipid lipid vesicles, zwitterionic phospholipids capable of a close-packing arrangement will prefer a position on the inner surface, compared with charged phospholipids which occupy a larger surface area and tend to orientate in the outer leaflet. Likewise, in vesicles consisting of a mixture of phospholipid and cholesterol, a higher proportion of the more condensed cholesterol in the inner leaflet could be explained by a substitution of cholesterol for phospholipid molecules with polyunsaturated hydrocarbon chains, which, because of steric hindrance, pack more loosely into the bilayer. Such interpretations also imply that asymmetry of phospholipids could be induced in biological membranes simply in response to changing membrane curvature. Moreover, segregation of lipid components in the plane of the membrane, as well as asymmetry across the bilayer, could be influenced by the folding of the membrane, thereby creating contiguous regions within membranes possessing highly specialised properties.

3.4.5 Lipid Asymmetry in Biological Membranes

Many situations are known in which lipids are required to confer biological activity on membrane-bound enzymes and surface antigens and, because most of these components are themselves arranged asymmetrically in the membrane, it may be inferred that the associated lipids are distributed accordingly. More direct evidence has been obtained for phospholipid:cholesterol asymmetry in myelin from X-ray electron-density profiles. Casper and Kirschner[71], for example, reported an approximately equimolar ratio of cholesterol and polar lipid in the outer side of the hydrocarbon layer of myelin but this ratio falls to about 3:7 on the inner side of the membrane. An asymmetric distribution of phospholipids across the erythrocyte membrane has also been demonstrated, but not much information is available yet on other biological membranes. Two types of approach have been adopted to investigate lipid asymmetry in erythrocyte membranes; the first employs the selective action of a number of specific phospholipase enzymes and the second involves the use of chemical reagents that react specifically with amino-containing phospholipids such as phosphatidylethanolamine and phosphatidylserine.

Some early studies of phospholipase action on erythrocytes from a number of species was reported by Turner and his colleagues[72,73]. They found that phospholipase A of cobra venom, an enzyme hydrolysing the fatty-acyl residues of glycerophosphatides, was lytic to most mammalian erythrocytes with the exception of those from ruminant species. The phospholipid composition of ruminant erythrocytes is distinct from those of other species because sphingomyelin largely replaces phosphatidylcholine as the predominant phospholipid class (see figure 1.14a) and this may explain, in part, their remarkable resistance to lysis. In contrast, when isolated erythrocyte membrane preparations from most species, including ruminants, are treated with phospholipase A they are rapidly depleted of all diacylglycerophosphatides because the enzyme is accessible to both sides of the membrane. This indicates that in ruminants at least sphingomyelin is the predominant phospholipid on the outer surface of the membrane. It should be noted that the proportion of diacylglycerophosphatides relative to plasmalogen in erythrocytes varies from one species to another and is high, for example, in human erythrocytes compared with those from ruminants. Enzymic procedures have been considerably refined by using highly purified phospholipase from different sources[74]. Pure phospholipase A_2, for example, which removes only the fatty-acyl residue attached to the carbon-2 position of glycerol, is not itself lytic to erythrocytes, but crude preparations of the enzyme often possess additional components that operate in conjunction with the enzyme to produce haemolysis. It is obvious that if the plasma membrane ruptures then enzyme action is not restricted solely to the outer surface of the membrane and no meaningful conclusions can be drawn with regard to phospholipid distribution.

Roelofsen *et al.*[75] have shown that pancreatic phospholipase A_2 and phospholipase C of *Bacillus cereus*, both of which attack glycerophosphatides but not sphingomyelin, are not lytic to human erythrocytes under isotonic conditions but neither do they produce any significant phospholipid breakdown. Furthermore, preliminary treatment of the cells with proteolytic enzymes or neuraminidase also fails to stimulate the hydrolytic activity of these enzymes. If the membrane is ruptured, however, and the cytoplasmic surface of the membrane is exposed to the enzymes, there is an almost complete degradation of the glycerophosphatides but, as expected, sphingomyelin remains intact. Other sources of phospholipase A_2 such as bee and cobra venoms are also non-haemolytic to human erythrocytes under isotonic conditions, but these enzymes hydrolyse about 20 per cent of the total phospholipids and nearly all of this is phosphatidylcholine. Quantitatively, this represents almost two-thirds of the total phosphatidylcholine, all of which is presumably located on the outer surface of the membrane. A rather interesting feature of these experiments is the fact that, even though substantial amounts of phosphatidylcholine are converted to lysophosphatidylcholine, the membrane appears to remain intact. This finding is rather remarkable considering that low concentrations of exogenous lysophosphatidylcholine (20 μM) cause membrane instability and fusion of a variety of cell types and produces haemolysis of red cells[76]. Roelofsen *et al.*[75] discovered that the susceptibility of membrane glycerophosphatides to hydrolysis by phospholipase enzymes was markedly dependent on the ionic strength of the suspending medium. When erythrocytes were treated with pancreatic phospholipase A_2 and *Bacillus cereus* phospholipase C under hypotonic conditions, for example, there was a rapid and extensive hydrolysis of phospholipids leading to haemolysis of the cells. This may indicate that penetration of the enzymes into the substrate is facilitated in some way by low salt concentration. In agreement with earlier observations, ruminant red cells were surprisingly resistant to haemolysis and no substantial breakdown of glycerophosphatides could be detected even when the salt concentration was reduced. This result is to be expected if these phospholipids are located predominantly on the inner surface of erythrocyte membranes. In other experiments[77] using a combination of phospholipases A_2 and C, it was found that resealed human erythrocyte membranes became leaky after digestion even under isotonic conditions, indicating that manipulation of the membrane associated with ghost preparation caused either a change in the susceptibility of glycerophosphatides to hydrolysis or an alteration of their disposition in the membrane.

Chemical-labelling procedures using reagents specific for amino groups has provided additional information regarding asymmetry of phosphatidylethanolamine and phosphatidylserine distribution in membranes. Bretscher's reagent, [^{35}S]-formylmethionyl sulphone methyl phosphate, (see section 3.4.3) for example, reacts only with a small proportion of the total phosphatidylethanolamine molecules of intact human erythrocytes, whereas most of this phospholipid becomes labelled in isolated membrane preparations. Similar results had been reported previously by Maddy[78] using the fluorescent reagent SITS (stilbene-4-acetamido-4′-thiocyanodisulphonate), which he showed reacts extensively with external proteins of intact bovine erythrocytes but not substantially with amino phospholipids. Whiteley and Berg[79] have synthesised a novel imidoester, isothionyl acetimidate, which cannot penetrate the human erythrocyte membrane but has the same reactivity towards amino groups as ethyl acetimidate, which does penetrate the membrane. Using ^{14}C and ^{3}H derivatives of the respective reagents they found that the cytoplasmic surface had ten times more reactive sites than the external surface of the membrane and that nearly all these sites were amino phospholipids. Furthermore, they showed that there was no drastic rearrangement of reactive groups on preparing isolated membrane fractions.

3.5 Membrane Biosynthesis and Turnover

Cell membranes must be synthesised, repaired or modified as required during the lifetime of the cell. There is now abundant evidence to show that a large number of membrane components are synthesised and degraded at a very rapid rate so that particular molecular species may turn over many times faster than the turnover rate or generation time of the cell. This implies that membrane structure is not static in the sense that a full complement of membrane components is required to preserve the integrity of a particular membrane at any one point in time. Moreover, turnover of this type is probably the main mechanism for introducing chemical modifications into membranes consistent with changes in their location or function. Studies of membrane biosynthesis of various cell types has therefore provided many useful insights into the dynamic aspects of membrane structure and the inter-relationships between different subcellular membranes. Nearly all mammalian cells, with the notable exception of erythrocytes, possess the machinery required to synthesise all membrane components necessary for their own growth and development. How these components are assembled into membranes is not fully understood but this could be by an assembly, spontaneous or otherwise, of all components into a piece of differentiated membrane or, just as one cell arises from another, one membrane may constitute the framework for the expansion and differentiation of new membrane.

The problem of finding out where and how membrane components are synthesised has been tackled mainly by following the fate of radioactively labelled precursors of various membrane proteins, lipids and carbohydrates. Thus when radioisotopically labelled phospholipid precursors or amino acids are injected into live animals, and the tissues subsequently analysed for the distribution of radioactive products, it is found that all cell membranes eventually become labelled. It is also apparent that membrane proteins must be synthesised at intracellular sites because it is difficult to imagine how proteins delivered from extracellular sources could penetrate the plasma membrane. Synthesis of lipids and carbohydrate components at extracellular sites, on the other hand, cannot be excluded on these grounds alone. It is known that the liver synthesises phospholipids in excess of its own requirements, but experiments with hepatectomised animals has largely excluded the possibility that phospholipids are synthesised by one organ and transported to another via the blood stream. It should be noted that the liver is the main source of plasma phospholipids, but the site of deposition of these lipids is somewhat obscure. Since all cell membranes are labelled when radioisotopic precursors are administered, the question arises as to whether these compounds are incorporated into membrane components at one or other site within the cell and transported through the cytoplasm to other membranes or whether each membrane synthesises its own phospholipid and protein requirements. The two main approaches to this problem have been, firstly, to locate the enzymes concerned with the synthesis of membrane components and, secondly, to conduct pulse-chase type experiments in which tissues are exposed to radioisotopically labelled precursors for a specified time interval, followed by an administration of a relatively large amount of the same unlabelled precursor. This procedure effectively dilutes the specific radioactivity so that the rate of appearance of labelled components in different subcellular membranes can be observed.

3.5.1 Synthetic Site of Membrane Proteins

Most membrane proteins appear to be synthesised on membrane-associated ribosomes of the rough-surface endoplasmic reticulum, on sites similar to those at which proteins destined for secretion from the cell are synthesised (see section 5.2). A few membrane proteins, however, are synthesised predominantly on free polyribosomes in the cytoplasm. One example is NADPH-cytochrome c reductase, an enzyme normally associated with the endoplasmic reticulum. Lowe and Hallinan[80] compared the incorporation of labelled amino acid into this enzyme in preparations of rough-surface endoplasmic reticulum and in free polyribosomes, and found, on average, a fourfold increase of synthesis in unbound ribosomes. Nevertheless, the rate of enzyme synthesis in some preparations of rough-surface membrane was appreciable, and they could not completely discount the possibility of a slight contamination of the membrane preparation with free polyribosomes or the attachment of these to the membrane by amino acid sequences nearing completion.

The site of synthesis of mitochondrial membrane proteins has proved to be of great interest. Studies of yeast have been particularly revealing because mitochondrial development in these organisms can be influenced to a large extent by their growth conditions. Yeast cells grown under anaerobic conditions and in the presence of a rapidly fermented sugar have few mitochondria and those mitochondria that are present lack certain enzymes of the electron-transport chain, such as cytochromes a, a_3, b, c and c_1, that are normally found in yeast provided with a non-fermentable substrate such as lactate. The fact that some mitochondrial proteins, but not others, are synthesised in catabolically repressed yeast raises the question of how the synthesis of these proteins is directed and controlled. Mitochondria synthesise about 10 per cent of their own protein, but no complete mitochrondrial enzyme is synthesised exclusively on mitochondrial ribosomes. Subunits of mitochondrial proteins must therefore be supplied from the cytoplasm or alternatively other cytoplasmic proteins are required to act in a regulatory manner. The mechanism of protein synthesis at the two sites differs in certain respects and those differences have been exploited to identify the site of synthesis of particular subunits of mitochondrial proteins.

A series of experiments in yeast has been described by Tzagoloff *et al.*[81] in which specific inhibitors of cytoplasmic and mitochondrial protein synthesis have been used to identify the site of synthesis of mitochondrial ATPase and cytochrome oxidase, both components of the inner mitochondrial membrane. These enzymes are oligomers of a number of non-identical subunits, which can readily be identified by gel electrophoresis in the presence of detergent. The incorporation of radioisotopically labelled amino acids into particular subunits was investigated in cells exposed to inhibitors of mitochondrial protein synthesis or cytoplasmic protein synthesis. Mitochondrial ATPase is known to consist of ten subunits; six of these are coded by nuclear genes and made on cytoplasmic ribosomes, whereas the remainder are synthesised on mitochondrial ribosomes and are presumably encoded in mitochondrial DNA. Similarly, cytochrome oxidase consists of seven subunits, four of which are of cytoplasmic origin and three of which are mitochondrial. It is noteworthy that the isolated subunits produced in the cytoplasm all tend to be water soluble, in contrast to the mitochondrial subunits, which are relatively insoluble peptides. Studies of protein synthesis in isolated mitochondria have shown that all three insoluble subunits of cytochrome oxidase are produced provided that oxygen is present; under anaerobic conditions only one subunit is synthesised, suggesting that oxygen may directly affect mitochondrial transcription or translation. Feldman and Mahler[82] used a different method to identify the site of protein synthesis. The initiator for mitochondrial protein synthesis is different from the cytoplasmic initiator and consists of formylmethionyl (f-Met)-tRNA so that yeast grown on media containing radioisotopically labelled formic acid incorporates labelled formate into proteins synthesised exclusively on mitochondrial ribosomes. These are nearly always associated with proteins of the inner mitochondrial membrane, but some labelled formate is incorporated in serine and finds its way into protein synthesised in the cytoplasm. Nevertheless, formyl peptides can be distinguished from labelled serine - containing peptides because the formyl group is acid labile. Feldman and Mahler went on to examine the relationship between protein synthesis at the two sites. They found that inhibition of cytoplasmic synthesis did not affect initiation of mitochondrial proteins, as judged by the incorporation of radioisotopically labelled formate into nascent peptide chains, but that the elaboration of complete membrane-associated protein ceased abruptly. This is consistent with the notion that the synthesis of mitochondrial and cytoplasmic components of mitochondrial proteins are tightly coupled and geared to the prevailing requirements of the cell.

Chloroplasts appear to be analogous to mitochondria because they too are capable of synthesising some of their own proteins independently of cytoplasmic protein synthesis. The major chloroplast protein, ribulose diphosphate carboxylase is, for example, composed of large subunits encoded by chloroblast DNA and small subunits encoded by nuclear genes.

Glycoproteins of the plasma membrane are another class of proteins that are synthesised at two sites. The peptide components of glycoproteins are thought to be produced on membrane-associated ribosomes of the rough-surface endoplasmic reticulum and carbohydrates are attached to these while *en route* to their ultimate destination on the surface of the plasma membrane. This scheme is consistent with studies conducted by Bosmann *et al.*[83] who investigated the rate of incorporation of radioisotopically labelled amino acid and carbohydrate into glycoproteins of smooth-surface internal membranes and the plasma membrane of cultured HeLa cells. The fact that almost no carbohydrate was recovered in the internal membranes of the cell, despite an appreciable incorporation of labelled amino acid into protein, suggested that the carbohydrate component was attached immediately prior to its incorporation into the plasma membrane. The site of carbohydrate incorporation has now been identified as the Golgi complex where trans-glycosylating enzymes required for these reactions are located.

3.5.2 Rate of Synthesis and Turnover of Membrane Proteins

In pulse-chase experiments with radioisotopically labelled amino acids, there is usually a long delay between synthesis of plasma-membrane proteins and their appearance on the cell surface. This can be seen in the experiments of Ray *et al.*[84] who isolated membranes from rat liver following a pulse of radioisotopically labelled amino acid, and measured the activity in different membranes (see figure 3.6). The activity of total tissue proteins remains relatively constant over a period of four to five hours. The distribution of radioactivity between smooth-surface endoplasmic reticulum and plasma-membrane fractions, however, changes in a manner suggesting that labelled proteins of the endoplasmic reticulum are decreasing and those of the plasma membrane are increasing by an equivalent amount. This is consistent with the fact that inhibiting protein synthesis with cycloheximide shortly after injecting the amino acid does not prevent a steady increase in the specific radioactivity of the plasma membrane.

Similar methods have been used by Barancik and Lieberman[85] to investigate changes in the specific radioactivity of different peptides isolated from liver plasma membranes at intervals after a pulse of radioisotopically labelled amino acid. They also used cycloheximide to prevent further protein synthesis, since this apparently does not affect the labelling pattern of membrane proteins already synthesised. They resolved at least eighteen peptides from the plasma membrane by gel electrophoresis and more than half of these had an apparent relative molecular mass greater than 50 000. The specific radioactivity of all membrane peptides increased during the first two hours following the radioactive amino acid pulse, but peptides of apparent relative molecular mass less than 50 000 had

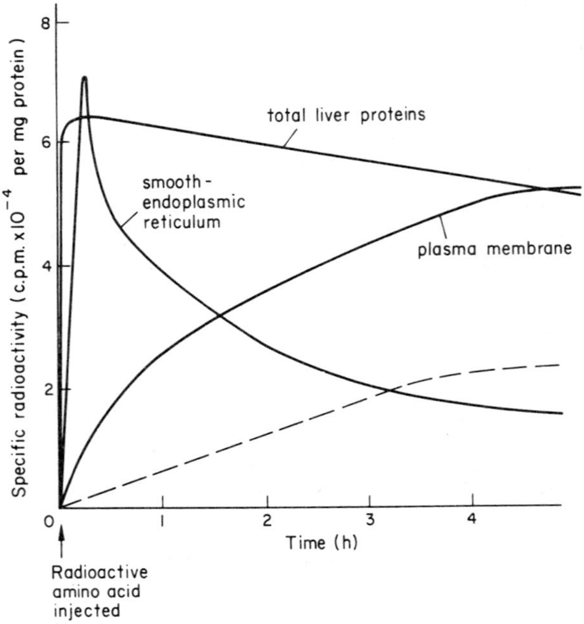

Figure 3.6 The incorporation of radioisotopically labelled amino acid into membranes of rat liver. The dashed line represents amino acid incorporated into plasma membranes of rats administered cycloheximide five minutes after the amino acid pulse (data from reference 84).

a higher specific radioactivity with fractions of higher relative molecular mass. This indicates that individual protein components are incorporated into the plasma membrane at different rates and possibly by different transporting mechanisms. It should be emphasised that these experiments were only concerned with the rate of incorporation of amino acid into peptides because, despite the tardy labelling of high relative molecular mass peptides, they are likely to have a shorter half-life; protein size is directly correlated with rate of degradation.

The rate of synthesis and degradation or turnover of proteins within particular membranes has also been shown to vary on the basis of the activity of various membrane-bound enzymes. Studies of the development and differentiation of the endoplasmic reticulum of embryonic rat and chick liver[86,87] have shown a continued fluctuation in the ratio of indigenous enzymes throughout development although, when examined under the electron microscope, the ultrastructure of the membrane appears to remain relatively unchanged. Many of these enzymes are functionally interdependent, so that their appearance at different stages of membrane differentiation is somewhat enigmatic. It is noteworthy that during development there is no marked change in the major phospholipid classes of the endoplasmic reticulum, apart from occasional differences in fatty-acyl distribution, which may simply reflect changes in the nutritional status of particular individuals.

Turnover studies of protein and lipid of smooth- and rough-surface membranes of rat liver have been reported by Omura *et al.*[88] in which membranes were analysed at intervals over eleven days following a single intraperitoneal injection of radioactive precursors. The decay of radioactivity in total protein and lipid extracts of each membrane indicated that lipids turn over at a significantly faster rate than proteins; the turnover rate of proteins was the same for smooth- and rough-surface membranes and the same was true for lipids. These observations were extended to compare the turnover rates of two membrane-specific enzymes – reduced NADP:ferricytochrome c oxidoreductase and cytochrome b_5. The two enzymes were extracted from preparations of each membrane and purified so that specific enzyme activity and radioactivity could be evaluated. It was found that the turnover rate of cytochrome b_5 was much slower than the reductase even though the former is synthesised predominantly on free polyribosomes (see section 3.5.1) and has a turnover rate closely approaching that of the average half-life of proteins of this membrane.

A number of interesting studies reported by Warren and Glick[89] indicate that the rate at which new components appear in the plasma membrane of mouse fibroblasts in tissue culture does not depend on the rate at which the cells are growing. This means that degradation or removal of membrane components from growing cells must be slower than in non-growing cells, in order to account for the net synthesis of membrane during growth and division of the cell. The mechanism of removal of components from the plasma membrane may be partially explained by the recovery of appreciable amounts of high relative molecular mass material from the medium surrounding non-growing, but not from growing, cells. It is possible that some of this material becomes detached from the surface of the plasma membrane of static cells. Other studies have been conducted on mouse and hamster tissue-culture cells to investigate the rate of appearance of membrane components at different stages in the cell growth cycle[90]. During the period immediately following one cell division and the commencement of the next there is a doubling of the total protein, lipid and carbohydrate of the plasma membrane. Protein and lipid components appeared to be inserted continuously throughout interphase although the incorporation of carbohydrate into the plasma membrane was variable. Certain specific-membrane components, such as 5′nucleotidase, (Mg^{2+})-ATPase, H-2 histocompatibility antigen and amino acid transport proteins, were all inserted during early interphase. The activity of other membrane-bound enzymes, however, varied through the cell cycle but enzyme inactivation or rapid turnover may have obscured the net synthesis of these components.

The rate at which plasma-membrane components are synthesised does not seem to vary even when cells are undergoing intense periods of activity. During phagocytosis, for example, when large areas of the plasma membrane are taken into the cell, there is apparently no increase in the rate of synthesis of plasma-membrane components compared with other internal membranes of the cell, such as the endoplasmic reticulum, or for that matter when compared with resting cells not undergoing phagocytosis. Moreover, Goodall *et al.*[91] demonstrated that plasma membrane removed from the cell surface during phagocytosis was replaced, even when protein synthesis was blocked, suggesting that there is an intracellular reserve of at least some plasma-membrane proteins. An interesting series of experiments relating to the intracellular reserves of glycoprotein-histocompatibility antigens has been reported recently by Nathenson and Cullen[92]. They treated intact mouse cells in tissue culture with papain to destroy antigens exposed on the surface of the cell and, after washing away the enzyme, observed the reappearance of antigens in the plasma membrane. New antigen begins to appear after a lag of about one hour and stops within six hours, by which time the number of membrane sites are restored to their original level. If protein synthesis is inhibited with cycloheximide, antigen reappears only up to two hours after papain treatment but subtle changes in the carbohydrate component of the new antigens was noted. (It is important to recognise that antigenic specificity of histocompatibility antigens resides in the protein component rather than in carbohydrates, as in the case, for example, of the blood-group antigens, where antigenic specificity ultimately relies on the activity of trans-glycosylating enzymes responsible for attaching the appropriate sugar residues.) Fucose and galactose incorporation into H-2 histocompatibility antigen was apparently normal in cycloheximide-treated cells but other sugars, such as galactosamine and mannose, that form the so-called core residues and are the first to be attached to the polypeptide chain, did not appear in new surface antigen. Since fucose and galactose are added last it seems likely that the most abundant intracellular reserve of surface antigen is in the form of partially completed glycoprotein molecules rather than as complete antigen or peptide precursors. Alternatively, inhibition of protein synthesis could interfere with the trans-glycosylating enzymes, which attach the core residues to the peptide precursor but do not affect enzymes attaching fucose or galactose.

The general conclusion from protein-turnover studies in animal cell membranes is that these components are inserted into an initial assembly of lipids and proteins that constitute what is tantamount to a primary undifferentiated membrane. It may be that even differentiated membranes are subject to modification or may even be transformed into different membranes during the normal course of cellular events.

3.5.3 Synthesis and Origin of Membrane Lipids

The predominant site of *de novo* synthesis of the major phospholipid classes is the endoplasmic reticulum. Enzymes involved in phosphatidylcholine synthesis, measured by the ability to incorporate radioisotopically labelled CDP-choline into phospholipid, are located in the endoplasmic reticulum of liver and rough- and smooth-surface membrane fractions, both of which appear to incorporate the precursor at about the same rate[93]. Mitochondria are also able to synthesise certain phospholipids, such as phosphatidic acid, cardiolipin and phosphatidylglycerol, but enzymes required for phosphatidylcholine synthesis from the precursor, CDP-choline or phosphatidylethanolamine from CDP-ethanolamine, are absent. Phospholipase A, however, has been found in the outer mitochondrial membrane so that a redistribution of fatty-acyl residues among different phospholipid molecules may be performed at this site. It is also important to recognise that, in intact cells, mitochondria provide the necessary energy to drive synthetic reactions taking place in the endoplasmic reticulum. Most of the final steps in the synthesis of phospholipids are performed by enzymes, such as choline- and ethanolamine–phosphotransferases, which are firmly membrane-bound, and since none of the products is freely water soluble it is assumed that they are incorporated directly into membranes or lipoproteins.

Mature erythrocytes are incapable of *de novo* synthesis of any membrane components and appear to rely on plasma sources, notably the plasma lipoproteins, for replenishment of membrane lipids. London and Schwarz[94] showed many years ago that erythrocytes were unable to synthesise cholesterol, yet there was an active turnover of unesterified cholesterol molecules in the membrane and from this they conceived a dynamic exchange of cholesterol between the membrane and blood plasma. Other workers subsequently showed that all the free cholesterol of the erythrocyte membrane is available for exchange and the replacement time of this cholesterol is of the order of eight to nine hours in mammalian cells. Phospholipids of the erythrocyte membrane are also likely to be supplied from circulating plasma lipoproteins[95], but some interconversions between phospholipid molecules probably take place once they become incorporated into the membrane. Proteins of the erythrocyte membrane turn over slowly in accordance with the lifetime of mature cells in circulation but, even so, the rate of turnover is faster than the total protein of the whole cell. Morrison *et al.*[96] observed the fate of erythrocyte membrane proteins iodinated by the lactoperoxidase method and found that, in contrast to other membranes, the lifetime of labelled membrane proteins was the same for all peptides irrespective of their relative molecular mass.

3.5.4 Turnover of Individual Phospholipid Classes

The basal turnover rate of most phospholipid classes in cell membranes is generally higher than that of proteins. The significance of this turnover is not fully appreciated but may result from the 'wear and tear' of membranes in general and play some role in membrane repair mechanisms. In addition, metabolic fluidity of phospholipids may confer a greater structural sensitivity to cell membranes with regard to hormone and other types of stimulation.

So far as the chemistry of phospholipid turnover is concerned, it is known that phospholipid classes differ in their susceptibility to phospholipase attack. Sphingomyelin, for example, is more stable than phosphatidylcholine, and turnover rates of sphingomyelin in membranes tend to be much slower. The nature of the acyl chain appears to be equally important in this respect; diacylglycerophosphatides with short unsaturated hydrocarbon chains are more susceptible to phospholipases A and C and turn over faster than molecules containing long unsaturated fatty acid groups. This difference may be due to the decreased hydrophobicity of phospholipids with short unsaturated chains, which allows them to take up a more exposed position on the membrane surface. Another factor that may influence susceptibility to phospholipase attack is the proximity of membrane proteins, particularly where specific interactions between phospholipids and proteins are involved.

In practice, the recommended procedure for assessing phospholipid turnover is to measure degradation of pre-labelled molecules rather than to determine the rate of incorporation of precursors, since this avoids complications due to fluctuations in the size and specific activities of precursor pools. Irrespective of whether synthesis or degradation is measured, it is important to distinguish between net turnover and isotope-exchange reactions, such as the transfer of residues from one molecule to another, since these may be of little consequence with regard to membrane structure or function.

Of all the major phospholipid classes, phosphatidic acid and phosphatidylinositol turn over in cell membranes at an appreciably faster rate than any other phospholipid. The metabolic lability of phosphatidic acid can be explained by the fact that it is a key intermediate in the synthesis of most other phospholipids but phosphatidylinositol, on the other hand, is quite exceptional. The synthetic pathway of this phospholipid is unique in the sense that inositol is attached to CDP-diglyceride. Furthermore, an enzyme is present in most tissues that specifically degrades phosphatidylinositol into diglyceride and inositol phosphates; the only other phospholipase C-type enzymes have been discovered in mammalian tissues and are specific for phosphatidylethanolamine and sphingomyelin.

An enhanced uptake of inorganic phosphate and other precursors into phosphatidylinositol and phosphatidic acid, compared with other phospholipid classes, has been known for some time to be associated with specific stimulation of a variety of different tissues. Thus, increased synthesis of phosphatidylinositol has been reported in polymorphonuclear leucocytes phagocytosing polystyrene granules and in lymphocytes stimulated with lectin. Hormone stimulation of target tissues elevates the incorporation of inorganic phosphate into phosphatidylinositol of adipose tissue stimulated by insulin, thyroid gland stimulated by pituitary thyrotrophin, pineal gland stimulated with a variety of phenethylamines and local anaesthetics and rat heart following adrenaline administration. Similar effects have been noted in nervous tissue treated with acetylcholine, low concentrations of certain tranquillising drugs, local anaesthetics and following electrical stimulation. The incorporation of precursors is, in many instances, found to reflect turnover rather than net synthesis of phosphatidylinositol because the total amount of the phospholipid remains relatively constant during and after stimulation. Moreover, phosphatidylinositol turnover is elevated within a few minutes of application of the primary stimulus, suggesting that synthesis or degradation of this phospholipid may be a prerequisite for subsequent cellular responses (see section 5.1.10).

External stimulation of cells is often mediated by elevation of intracellular cyclic AMP concentrations and there is some evidence that phosphatidylinositol is associated with membrane-bound adenylcyclase and influences hormone-binding to the receptor site of the enzyme (see section 5.1.3). Although many hormones stimulate phosphatidylinositol turnover, this effect is not usually duplicated by cyclic AMP or its dibutyryl analogue despite the fact that these compounds mimic the normal physiological response. It seems, therefore, that the coupling between phosphatidylinositol turnover and cellular response may not be as direct as was originally thought. Furthermore, β-adrenergic blocking agents, local anaesthetics and phenothiazines abolish hormonal response in some tissues without diminishing phosphatidylinositol turnover. Some of these agents possess surface-active properties and this could be the mechanism for initiating phospholipid turnover.

3.5.5. Phospholipid Exchange Between Membranes

Isotopic experiments have provided convincing evidence that phospholipids of all membranes turn over at a relatively rapid rate, albeit much faster in some membranes, like the endoplasmic reticulum, than in others, such as myelin. Since phospholipids are synthesised mainly in the endoplasmic reticulum, it is clear that intact phospholipid molecules must be transported from this site and inserted into other membranes located elsewhere in the cell. The participation of a soluble protein in phospholipid transfer reactions has been reported by a number of different workers[97,98,99]. In all these studies preparations of endoplasmic reticulum from rat liver, containing radioisotopically labelled phospholipids, were added to a suspension of unlabelled mitochondria and the appearance of labelled phospholipid in the mitochondria was observed. The addition of a soluble liver cell extract greatly facilitated the transfer of phospholipid from endoplasmic reticulum to mitochondria and, since the factor was heat-labile and non-dialysable, it was inferred that a protein was responsible. It was also found that labelled phospholipid could be transferred with equal facility from mitochondria to endoplasmic reticulum and even from aqueous dispersions of phospholipid to membranes, suggesting that the transfer was accomplished by an exchange-type reaction. There is one report, however, indicating a net transfer of phospholipid from liposomes to membranes, but the participation of phospholipid-exchange reactions in net membrane synthesis has yet to be verified.

Specific phospholipid-exchange proteins appear to be required for each of the major phospholipid classes and the protein responsible for phosphatidylcholine exchange has been identified in many mammalian tissues including liver, heart, kidney and brain. Kamp *et al.*[100] have purified this protein from beef liver and showed that it catalysed a rapid exchange of phosphatidylcholine from endoplasmic reticulum to mitochondria, but other phospholipids like phosphatidylethanolamine and phosphatidylinositol were not transferred. In experiments with monomolecular films of phosphatidylcholine[101] it has been shown that phosphatidylcholine-exchange protein transfers phospholipid between membrane interfaces and that each molecule of exchange protein is always associated with one molecule of phosphatidylcholine.

3.6 Summary

All cell membranes consist of lipids and proteins, which associate together or interact with one another in the membrane. The lipids occupy a mainly separate domain in a bimolecular arrangement characteristic of phospholipids dispersed in aqueous media. Proteins are not indispensible to the structure of membranes but are supported and organised within the bilayer matrix. The extent to which proteins interact with lipids ranges from adsorption – largely by polar forces – to the surface of the lipid bilayer, to an almost complete interpolation into the hydrocarbon region of the membrane. Some membrane proteins completely interrupt the bilayer, spanning the membrane from one polar surface to the other.

Membranes are thought to consist of a mosaic of lipid bilayer and proteins. Some regions of the membrane are believed to be fluid with molecules able to diffuse freely in the plane of the membrane while the relative motion of individual molecules in other regions may be more constrained. Apart from specific interactions between oligomeric units of multi-enzyme systems, membrane proteins are usually distributed randomly as individual molecules in the plane of the membrane. This has been confirmed by treating membranes with bifunctional cross-linking reagents. Glycoproteins of the erythrocyte membrane, for example, do not normally associate with other membrane proteins in the same plane of the membrane but their distribution appears to be important in many cell-surface responses. This distribution is maintained by thermal motion and charge repulsion between ionised polar groups. Such motion is reflected in a rapid rotation of protein molecules about an axis perpendicular to the plane of the membrane and at a rate depending on the viscosity or fluidity of the surrounding membrane matrix. The lateral mobility of membrane proteins is evident from their ability to aggregate when electrostatic repulsion is modulated or specific binding sites are complexed with multivalent lectins or antibodies. Diffusion of species-specific antigens over the entire surface of heterokaryon cells has been demonstrated. The rate at which an almost complete intermixing of these antigens is achieved depends on the fluidity of the lipid matrix and does not require metabolic energy. When lectin or antibody binding sites are complexed on the surface of certain cell types, such as lymphocytes, they aggregate into patches at several points on the cell periphery. These patches are directed towards one or other pole of the cell and may be internalised by endocytosis of the plasma membrane. Unlike patch formation, the direction of complexed antigen into a polar cap requires metabolic energy as well as a fluid lipid matrix.

Some selection of membrane components, internalised when plasma membrane is removed from the cell surface during phagocytosis, is likely. Reagents that disrupt cytoplasmic microtubular and microfilamentous structures prevent discrimination, suggesting a possible connection between cytoplasmic structures and specific membrane proteins.

The interaction between cells appears to be restricted to specialised regions of the cell surface. Depending on the nature of the components sequestered in these regions, membrane contacts may be unstable and the two membranes will fuse, or stable contacts may be established between adjacent membranes. These so-called intercellular junctions provide cohesion between cells, while other types of junction facilitate communication between connecting cells.

Protein, lipid and carbohydrate components are asymmetrically arranged in membranes. Except for Golgi vesicles, carbohydrate components are presented only on the outer surface of the plasma membrane where they form a so-called cell coat. Proteins labelled with impermeable reagents or peptides cleaved from the surface of intact cells with proteases are distinctly different from those accessible on the cytoplasmic surface of the membrane. Similarly, phospholipids susceptible to hydrolysis by specific phospholipase enzymes or labelled with impermeable reagents differ on each side of the membrane. Molecular asymmetry is germane to the conduct of enzyme reactions at particular locations on the membrane surface as well as the presentation of specific receptor sites and antigens on the cell surface but the reason for lipid asymmetry is not yet clear.

Nearly all cells can synthesise all the membrane components necessary for their growth and development. The endoplasmic reticulum is the predominant site of membrane protein and phospholipid synthesis, although mitochondria possess a limited capacity to synthesise some of these components. Carbohydrates are attached to peptide precursors of glycoproteins in the Golgi apparatus. Membrane proteins and lipids turn over at different rates in membranes. The rate of synthesis appears to be independent of cell growth or membrane consumption during cellular activity, so that the rate of loss of individual components from membranes is the main variable in turnover equations. Phospholipids turn over at a significantly faster rate than proteins and the turnover rate varies between phospholipid classes; phosphatidylinositol and phosphatidic acid are appreciably more labile than the other major phospholipid classes. Phospholipids can be transferred between membranes by specific phospholipid-exchange proteins but it is doubtful whether such exchange reactions contribute to net synthesis of membrane.

The dynamic structure of membranes is exemplified by the fact that individual components can be inserted and removed, possibly at different locations, without perturbing the overall structure of the membrane. Nearly all membranes are assembled by a multistep process which permits the differentiation of new membrane from pre-existing membrane structures.

References

1. E. Gorter and F. Grendel. Bimolecular layers of lipoids on chromatocytes of blood. *J. exp. Med.*, **41** (1925), 439-43
2. E. Overton. Über die osmotischen Eigenschafler der lebenden Pflanzen und Tierzelle. *Vjschr. naturf. Ges. Zürich,* **40** (1895), 159-201
3. I. Langmuir. Molecular films in chemistry and biology. In: *Molecular Films, the Cyclotron and the New Biology,* Essays by H. S. Taylor *et al.,* Rutgers University Press, New Brunswick (1942), pp 27-62
4. R. S. Bar, D. W. Deamer and D. G. Cornwell. Surface area of human erythrocyte lipids: reinvestigation of experiments on plasma membrane. *Science, N. Y.,* **153** (1966), 1010-12
5. D. M. Engelman. Surface area per lipid molecule in the intact membrane of the human red cell. *Nature, Lond.,* **223** (1969), 1279-80
6. A. R. Limbrick and S. Knutton. A structural study of the modification of erythrocyte membranes by phospholipase C. *Abs. XI A. Meeting Am. Soc. Cell Biol.,* (1971), p. 168
7. J. F. Danielli and H. Davson. A contribution to the theory of permeability of thin films. *J. cell. comp. Physiol.,* **5** (1935), 495-508
8. J. D. Robertson. *The Ultrastructure of Cell Membranes and their Derivatives.* Biochem. Soc. Symposia No 16, Cambridge University Press, London (1959), pp 3-43

9. W. J. Dreyer, D. S. Papermaster and H. Kühn. On the absence of ubiquitous structural protein subunits in biological membranes. *Ann. N.Y. Acad. Sci.*, **195** (1972), 61–74

10. S. J. Singer and G. L. Nicolson. The fluid mosaic model of the structure of cell membranes. *Science, N.Y.*, **175** (1972), 720–31

11. M. Edidin. Rotational and translational diffusion in membranes. *A. Rev. Biophys. Bioeng.*, **3** (1974), 179–201

12. C. J. Scandella, P. Devaux and H. M. McConnell. Rapid lateral diffusion of phospholipids in rabbit sarcoplasmic reticulum. *Proc. natn. Acad. Sci. U.S.A.*, **69** (1972), 2056-60

13. A. J. Verkleij, P. H. J. Th. Ververgaert, L. L. M. Van Deenen and P. F. Elbers. Phase transitions of phospholipid bilayers and membranes of *Acholeplasma laidlawii B* visualised by freeze-fracturing electron microscopy. *Biochim. Biophys. Acta*, **288** (1972), 326–32

14. F. Wunderlich, R. Müller and V. Speth. Direct evidence for a colchicine-induced impairment in the mobility of membrane components. *Science, N.Y.*, **182** (1973), 1136–8

15. J. Torres-Pereira, R. Mehlhorn, A. D. Keith and L. Packer. Changes in membrane lipid structure of illuminated chloroplasts – studies with spin-labeled and freeze-fractured membranes. *Archs Biochem. Biophys.*, **160** (1974), 90–9

16. P. K. Brown. Rhodopsin rotates in the visual receptor membrane. *Nature new Biol.*, **236** (1972), 35–8

17. R. A. Cone. Rotational diffusion of rhodopsin in the visual receptor membrane. *Nature new Biol.*, **236** (1972), 39–43

18. P. Pinto da Silva. Translational mobility of the membrane intercallated particles of human erythrocyte membranes. pH-dependent reversible aggregation. *J. Cell Biol.*, **53** (1972), 777–87

19. P. Pinto da Silva, P. S. Moss and H. H. Fudenberg. Anionic sites on the membrane intercallated particles of human erythrocyte ghost membranes freeze–etch localization. *Expl. Cell Res.*, **81** (1973), 127–38

20. G. L. Nicolson and R. G. Painter. Anionic sites of human erythrocyte membranes. II Antispectrin induced transmembrane aggregation of the binding sites for positively charged colloidal particles. *J. Cell Biol.*, **59** (1973), 395–406

21. H. R. Bose and M. A. Brundige. Selective association of sindus virion proteins with different membrane fractions of infected cells. *J. Virol.*, **9** (1972), 785–91

22. C. Guerin, A. Zachowski, B. Prigent, A. Paraf, I. Dunia, M-A. Diawara and E. L. Benedetti. Correlation between the mobility of inner plasma membrane structure and agglutination by concanavalin A in two cell lines of MOPC 173 plastocytoma cells. *Proc. natn. Acad. Sci. U.S.A.*, **71** (1974), 114–17

23. J. Z. Rosenblith, T. E. Ukena, H. H. Yin, R. D. Berlin and M. J. Karnovsky.

A comparative evaluation of the distribution of concanavalin A-binding sites on the surfaces of normal, virally transformed, and protease-treated fibroblasts. *Proc. natn. Acad. Sci. U.S.A.*, **70** (1973), 1625–9

24. G. R. Gunther, J. L. Wang, I. Yahara, B. A. Cunningham and G. M. Edelman. Concanavalin A derivatives with altered biological activities. *Proc. natn. Acad. Sci. U.S.A.*, **70** (1973), 1012–16
25. L. D. Frye and M. Edidin. The rapid intermixing of cell surface antigens after formation of mouse–human heterokaryons. *J. Cell Sci.*, **7** (1970), 319–35
26. R. B. Taylor, W. P. H. Duffus, M. C. Raff and S. de Petris. Redistribution and pinocytosis of lymphocyte surface. Immunoglobulin molecules induced by anti-immunoglobulin antibody. *Nature new Biol.*, **233** (1971), 225–9
27. S. de Petris and M. C. Raff. Distribution of immunoglobulin on the surface of mouse lymphoid cells as determined by immunoferritin electron microscopy. Antibody-induced, temperature-dependent redistribution and its implications for membrane structure. *Eur. J. Immunol.*, **2** (1972), 523–35
28. S. de Petris and M. C. Raff. Normal distribution, patching and capping of lymphocyte surface immunoglobulin studied by electron microscopy. *Nature new Biol.*, **241** (1973), 257–9
29. C. W. Stackpole, J. B. Jacobson and M. P. Lardis. Two distinct types of capping of surface receptors on mouse lymphoid cells. *Nature, Lond.*, **248** (1974), 232–4
30. M. F. Tsan and R. D. Berlin. Effect of phagocytosis on membrane transport of non-electrolytes. *J. exp. Med.*, **134** (1971), 1016–35
31. J. M. Oliver, T. E. Ukena and R. D. Berlin. Effects of phagocytosis and colchicine on the distribution of lectin-binding sites on cell surfaces. *Proc. natn. Acad. Sci. U.S.A.*, **71** (1974), 394–8
32. T. E. Ukena and R. D. Berlin. Effect of colchicine and vinblastine on the topographical separation of membrane functions. *J. exp. Med.*, **136** (1972), 1–7
33. G. M. Edelman, I. Yahara and J. L. Wang. Receptor mobility and receptor-cytoplasmic interactions in lymphocytes. *Proc. natn. Acad. Sci. U.S.A.*, **70** (1973), 1442–6
34. R. D. Berlin, J. M. Oliver, T. E. Ukena and H. H. Yin. Control of cell surface topography. *Nature, Lond.*, **247** (1974), 45–6
35. T. L. Steck. Cross-linking the major proteins of the isolated erythrocyte membrane. *J. molec. Biol.*, **66** (1972), 295–305
36. R. A. Capaldi. A cross-linking study of the beef erythrocyte membrane: extensive interaction of all the proteins of the membrane except for the glycoproteins. *Biochem. Biophys. Res. Commun.*, **50** (1973), 656–61
37. T. H. Ji. Crosslinking of the glycoproteins in hyman erythrocyte membranes. *Proc. natn. Acad. Sci. U.S.A.*, **71** (1974), 93–5
38. G. V. Marinetti, D. S. Sheeley, R. Baumgarten and R. Love. Cross-linking of phospholipid neighbours in the erythrocyte membrane. *Biochem.*

Biophys. Res. Commun., **59** (1974), 502–7

39. G. M. Edelman and C. F. Millette. Molecular probes of spermatozoan structure. *Proc. natn. Acad. Sci. U.S.A.*, **68** (1971), 2436–40

40. T. Oikawa, R. Yanagimachi and G. L. Nicolson. Wheat germ agglutinin blocks mammalian fertilization. *Nature, Lond.*, **241** (1973), 256–9

41. G. C. Koo, C. W. Stackpole, E. A. Boyse, U. Hämmerling and M. P. Lardis. Topographical location of H–Y antigen on mouse spermatozoa by immunoelectronmicroscopy. *Proc. natn. Acad. Sci. U.S.A.*, **70** (1973), 1502–5

42. H. Bittinger and H. P. Schnebli. Binding of concanavalin A and ricin to synaptic junctions of rat brain. *Nature, Lond.*, **249** (1974), 370–1

43. W. R. Loewenstein. Membrane junctions in growth and differentiation. *Fedn Proc. Fedn Am. Socs exp. Biol.*, **32** (1973), 60–4

44. N. S. McNutt and R. S. Weinstein. Membrane ultrastructure at mammalian intercellular junctions. *Prog. Biophys. molec. Biol.*, **26** (1973), 45–101

45. J. B. Wade and M. J. Karnovsky. The structure of the zona occludens. A single fibril model based on freeze–fracture. *J. Cell Biol.*, **60** (1974), 168–80

46. A. R. Muir. The effects of divalent cations on the ultrastructure of the perfused rat heart. *J. Anat.*, **101** (1967), 239–61

47. M. V. L. Bennett. Function of electrotonic junctions in embryonic and adult tissues. *Fedn Proc. Fedn Am. Socs exp. Biol.*, **32** (1973), 65–75

48. N. B. Gilula, O. R. Reeves and A. Steinbach. Metabolic coupling, ionic coupling and cell contacts. *Nature, Lond.*, **235** (1972), 262–5

49. R. Azarnia, W. J. Larsen and W. R. Loewenstein. The membrane junctions in communicating and non-communicating cells, their hybrids and segregants. *Proc. natn. Acad. Sci. U.S.A.*, **71** (1974), 880–4

50. N. S. McNutt and R. S. Weinstein. The ultrastructure of the nexus. A correlated thin section and freeze–cleave study. *J. Cell Biol.*, **47** (1970), 666–88

51. D. A. Goodenough and W. Stoeckenius. The isolation of mouse hepatocyte gap junctions. Preliminary chemical characterization and X-ray diffraction. *J. Cell Biol.*, **54** (1972), 646–56

52. W. H. Evans and J. W. Gurd. Preparation and properties of nexuses and lipid-enriched vesicles from mouse liver plasma membranes. *Biochem. J.*, **128** (1972), 691–700

53. G. L. Nicolson and S. J. Singer. Ferritin-conjugated plant agglutinins as specific saccharide stains for electron microscopy: application to saccharides bound to cell membranes. *Proc. natn. Acad. Sci. U.S.A.*, **68** (1971), 942–5

54. A. Rambourg. Morphological and histochemical aspects of glycoproteins at the surface of animal cells. *Int. Rev. Cytol.*, **31** (1971), 57–114

55. D. F. H. Wallach and E. H. Eylar. Sialic acid in the cellular membranes of Ehrlich ascites-carcinoma cells. *Biochim. Biophys. Acta*, **52** (1961), 594–6

56. G. Gasic and T. Gasic. Removal and regeneration of the cell coating in tumor cells. *Nature, Lond.*, **196** (1962), 170

57. D. F. H. Wallach. The dispositions of proteins in the plasma membranes of animal cells. Analytical approaches using controlled peptidolysis and protein labels. *Biochim. Biophys. Acta,* **265** (1972), 61–83

58. T. L. Steck, G. Fairbanks and D. F. H. Wallach. Disposition of the major proteins in the isolated erythrocyte membrane. Proteolytic dissection. *Biochemistry,* **10** (1971), 2617–24

59. M. S. Bretscher. A major protein which spans the human erythrocyte membrane. *J. molec. Biol.,* **59** (1971), 351–7

60. W. W. Bender, H. Garan and H. C. Berg. Proteins of the human erythrocyte membrane as modified by pronase. *J. molec. Biol.,* **58** (1971), 783–97

61. D. B. Rifkin, R. W. Compans and E. Reich. A specific labelling procedure for proteins on the outer surface of membranes. *J. biol. Chem.,* **247** (1972), 6432–7

62. O. O. Blumenfeld, P. M. Gallop and T. H. Liao. Modification and introduction of a specific radioactive label into the erythrocyte membrane sialoglycoproteins. *Biochem. Biophys. Res. Commun.,* **48** (1972), 242–51

63. R. Schmidt-Ullrich, H. Knüfermann and D. F. H. Wallach. The reaction of 1-dimethylaminonaphthalene-5-sulphonyl chloride (DANSCR) with erythrocyte membranes. A new look at 'vectorial' membrane probes. *Biochim. Biophys. Acta,* **307** (1973), 353–65

64. C. G. Gahmberg and S. Hakomori. External labelling of the cell surface galactose and galactosamine in glycolipid and glycoprotein of human erythrocytes. *J. biol. Chem.,* **248** (1973), 4311–17

65. D. R. Phillips and M. Morrison. The arrangement of proteins in the human erythrocyte membrane. *Biochem. Biophys. Res. Commun.,* **40** (1970), 284–9

66. E. Reichstein and R. Blostein. Asymmetric iodination of the human erythrocyte membrane. *Biochem. Biophys. Res. Commun.,* **54** (1973), 494–500

67. D. R. Phillips and M. Morrison. Changes in accessibility of plasma membrane protein as the result of tryptic hydrolysis. *Nature new Biol.,* **242** (1973), 213–15

68. C. T. Huber and M. Morrison. Heterogeneity of the outer membrane of mitochondria. *Biochemistry,* **12** (1973), 4274–82

69. D. M. Michaelson, A. F. Horwitz and M. P. Klein. Transbilayer asymmetry and surface homogeneity of mixed phospholipids in co-sonicated vesicles. *Biochemistry,* **12** (1973), 2637–45

70. C-H. Huang, J. P. Sipe, S. T. Chow and R. B. Martin. Differential interaction of cholesterol with phosphatidylcholine on the inner and outer surfaces of lipid bilayer vesicles. *Proc. natn. Acad. Sci. U.S.A.,* **71** (1974), 359–62

71. D. L. D. Casper and D. A. Kirschner. Myelin membrane structure at 10 Å resolution. *Nature new Biol.,* **231** (1971), 46–52

72. J. C. Turner. Absence of lecithin from the stromata of the red cells of certain animals (ruminants) and its relation to venom hydrolysis. *J. exp.*

Med., **105** (1957), 189-93

73. J. C. Turner, H. M. Anderson and C. P. Gandal. Species differences in red blood cell phosphatides separated by column and paper chromatography. *Biochim. Biophys. Acta*, **30** (1958), 130-4
74. R. F. A. Zwaal, B. Roelofsen and C. M. Colley. Localization of red cell membrane constituents. *Biochim. Biophys. Acta*, **300** (1973), 159-82
75. B. Roelofsen, R. F. A. Zwaal, P. Comfurius, C. B. Woodward and L. L. M. Van Deenen. Action of pure phospholipase A_2 and phospholipase C on human erythrocytes and ghosts. *Biochim. Biophys. Acta*, **241** (1971), 925-9
76. J. A. Lucy. The fusion of biological membranes. *Nature, Lond.*, **227** (1970), 814-17
77. C. B. Woodward and R. F. A. Zwaal. The lytic behaviour of pure phospholipases A_2 and C towards osmotically swollen erythrocytes and resealed ghosts. *Biochim. Biophys. Acta*, **274** (1972), 272-8
78. A. H. Maddy. A fluorescent label for the outer components of the plasma membrane. *Biochim. Biophys. Acta*, **88** (1964), 390-9
79. N. M. Whiteley and H. C. Berg. Amidination of the outer and inner surfaces of the human erythrocyte membrane. *J. molec. Biol.*, **87** (1974), 541-61
80. D. Lowe and T. Hallinan. Preferential synthesis of a membrane-associated protein by free polyribosomes. *Biochem. J.*, **136** (1973), 825-8
81. A. Tzagoloff, M. S. Rubin and M. F. Sierra. Biosynthesis of mitochondrial enzymes. *Biochim. Biophys. Acta*, **301** (1973), 71-104
82. F. Feldman and H. R. Mahler. Mitochondrial biogenesis. Retention of terminal formylmethionine in membrane proteins and regulation of their synthesis. *J. biol. Chem.*, **249** (1974), 3702-9
83. H. B. Bosmann, A. Hagopian and E. H. Eylar. Cellular membranes: the biosynthesis of glycoprotein and glycolipid in HeLa cell membranes. *Archs Biochem. Biophys.*, **130** (1969), 573-83
84. T. K. Ray, I. Lieberman and A. I. Lansing. Synthesis of plasma membrane of the liver cell. *Biochem. Biophys. Res. Commun.*, **31** (1968), 54-8
85. L. C. Barancik and I. Lieberman. The kinetics of incorporation of protein into liver plasma membrane. *Biochem. Biophys. Res. Commun.*, **44** (1971), 1084-8
86. G. Dallner, P. Siekevitz and G. E. Palade. Biogenesis of endoplasmic reticulum membranes: II. Synthesis of constitutive microsomal enzymes in developing rat hepatocyte. *J. Cell Biol.*, **30** (1966), 97-117
87. J. K. Pollak and D. B. Ward. Changes in the chemical composition and the enzymic activities of hepatic microsomes of the chick embryo during development. *Biochem. J.*, **103** (1967), 730-8
88. T. Omura, P. Siekevitz and G. E. Palade. Turnover of constituents of the endoplasmic reticulum membranes of rat hepatocytes. *J. biol. Chem.*, **242** (1967), 2389-96

89. L. Warren and M. C. Glick. Membranes of animal cells: II. The metabolism and turnover of the surface membrane. *J. Cell Biol.*, **37** (1968), 729-46

90. J. M. Graham, M. C. B. Sumner, D. H. Curtis and C. A. Pasternak. Sequence of events in plasma membrane assembly during the cell cycle. *Nature, Lond.*, **246** (1973), 291-5

91. R. J. Goodall, Y. F. Lai and J. E. Thompson. Turnover of plasma membrane during phagocytosis. *J. Cell Sci.*, **11** (1972), 569-79

92. S. G. Nathenson and S. E. Cullen. Biochemical properties and immunochemical-genetic relationships of mouse H-2 alloantigens. *Biochim. Biophys. Acta*, **344** (1974), 1-25

93. W. C. Schneider. Intracellular distribution of enzymes XIII. Enzymic synthesis of deoxycytidine diphosphate choline and lecithin in rat liver. *J. biol. Chem.*, **238** (1963), 3572-8

94. I. M. London and H. Schwarz. Erythrocyte metabolism. The metabolic behaviour of the cholesterol of human erythrocytes. *J. clin. Invest.*, **32** (1953), 1248-52

95. R. M. C. Dawson. The metabolism of animal phospholipids and their turnover in cell membranes. *Essays Biochem.*, 2 (1966), 69-115

96. M. Morrison, A. W. Michaels, D. R. Phillips and S. L. Choi. Life span of erythrocyte membrane protein. *Nature, Lond.*, **248** (1974), 763-89

97. K. W. A. Wirtz and D. B. Zilversmit. Exchange of phospholipids between liver mitochondria and microsomes *in vitro*. *J. biol. Chem.*, **243** (1968), 3596-602

98. W. C. McMurray and R. M. C. Dawson. Phospholipid exchange reactions within the liver cell. *Biochem. J.*, **112** (1969), 91-108

99. M. Akiyama and T. Sakagami. Exchange of mitochondrial lecithin and cephalin with those in rat liver microsomes. *Biochim. Biophys. Acta*, **187** (1969), 105-12

100. H. H. Kamp, K. W. A. Wirtz and L. L. M. Van Deenen. Some properties of phosphatidylcholine exchange protein purified from beef liver. *Biochim. Biophys. Acta*, **318** (1973), 313-25

101. R. A. Demel, K. W. A. Wirtz, H. H. Kamp, W. S. M. Geurts van Kessel and L. L. M. Van Deenen. Phosphatidylcholine exchange protein from beef liver. *Nature new Biol.*, **246** (1973), 102-5

4 Membrane Permeability and Transport

The chemical content of living cells is usually very different from that of its surroundings and this is maintained by the limiting membrane that bounds each cell. We know that, in order to sustain life, substrates need to be supplied from extracellular sources and must be afforded access through this membrane. Conversely, provision must be made for the egress of metabolic waste products, which would otherwise accumulate in toxic amounts to the detriment of the cell. Furthermore in view of the permeability of cell membranes to water the intracellular concentration of ions must be carefully controlled in the absence of a rigid cell wall in order to regulate cell volume and prevent a build up of hydrostatic pressure which could rupture the membrane. These rather perfunctory observations serve to illustrate that the membrane surrounding cells allows the penetration of some substances but not others; that is to say the membrane is selectively permeable.

The exchange of substances between individual cells and extracellular media takes place at the level of the plasma membrane, but in higher organisms specialised epithelial tissues constitute the primary interface for exchange of substances between the body and its environment. These tissues, which include intestinal mucosa, lung alveoli, renal tubules and skin, consist of one or more layers of cells but ultimately it is the particular properties of the membrane surrounding each individual cell that regulate the passage of substances across epithelia. There are two main pathways by which substances can be transported across an epithelium. The first is through the extracellular space between cells but, as we have already seen, movement by this route is impeded by *zona occludens* type intercellular junctions, which bind the cells tightly together. The other route is via a transcellular pathway by which molecules penetrate the plasma membrane on one side of the cell, diffuse through the cytoplasm and finally escape across the plasma membrane on the opposite side of the cell. A variation of this pathway has been suggested whereby substances are taken into the cell by endocytosis on one side of the cell and remain isolated from the cytoplasmic pool by enclosure in membrane vesicles during transport to the opposite side of the cell, where they are released to extracellular space by exocytosis. This mechanism is obviously restricted to highly impermeative solutes and is not likely to constitute a major transport pathway across epithelia. Total conductance across epithelia may therefore be described as the sum of the conductances of the transcellular and extracellular pathways and in practice it has been found that the relative conductance varies from one type of epithelium to another as well as depending on the physical properties of the particular solute considered.[1]

One reason for this difference is that the permeability characteristics of membranes differ from cell to cell, and in many instances permeability varies from one region to another in the same membrane. We have already considered electrically coupled gap junctions in this connection and found these to be highly permeable regions of the plasma membrane with specialised features quite distinct from other regions of the same membrane. Other subtle differences in membrane permeability around the periphery of some cells have been noted. Epithelial cells, for example, are known to be highly permeable to inorganic cations on the serosal side but virtually impermeable to these ions on the mucosal side. Furthermore the membranes of nerves and other excitable cells possess highly specialised permeability characteristics, which enable them to undergo very rapid changes in permeability whereby the free energy stored as gradients of sodium and potassium can be utilised to generate an electric current. This breakdown of membrane resistance is considered to be the primary event in the production of an action potential. It is apparent that the permeability characteristics of cell membranes play an important part in the normal maintenance of cellular activity and are essential to the performance of a number of physiological functions. We shall now address ourselves to the question of how different molecules pass across cell membranes; however, since most studies have been confined to transport across the plasma membrane, this membrane will figure more prominently. Substrate transport across the inner mitochondrial membrane is considered separately in a later section (5.3.2).

4.1 Diffusion of Solutes through Membranes

If molecules pass through a membrane at a rate directly proportional to the concentration gradient of those particular molecules across the membrane, the simplest explanation to account for their transport is by diffusion, according to some adaptation of Fick's first law. Ideally the process is independent of the magnitude of the solute gradient across the membrane, in contrast to other transporting mechanisms that require specific interactions between permeant and carrier and can be saturated at a finite concentration of solute molecules. Thus the passive diffusion of most molecules can be observed if the gradient across the membrane is in excess of that required to saturate other transporting systems.

Two theories have been proposed to account for the mechanism of passive transport of solutes across membranes. The first, known as the energy-barrier hypothesis, envisages an energy barrier, possibly sited in the interfacial region separating the aqueous from the hydrocarbon region of the membrane, across which permeant molecules must pass. In other words, penetrating molecules are required to strike the membrane with sufficient kinetic energy to force a hole in the membrane, through which they can pass. The alternative process is by simple diffusion according to Fick's first law, which assumes that the water–hydrocarbon interface is stable and the rate of solute transfer is governed solely by the solubility of the solute and its diffusion characteristics in the hydrocarbon region of the membrane. This mechanism implies that permeant solutes become physically dissolved in the membrane and move by random molecular–molecular motion through the lattice. The mathematical expression of the rate of flux J of a solute across a membrane is

$$J = -\frac{D\beta}{l}(C_1 - C_2) \tag{4.1}$$

where D and β are the diffusion and partition coefficients, respectively, of a solute whose concentration is C_1 and C_2 on either side of a membrane of thickness (hydrocarbon region) l. The negative sign of the equation accords recognition to the fact that solutes move in a direction of decreasing concentration. The term $D\beta/l$ is referred to as the permeability coefficient (P) of the membrane. The permeability coefficient can be measured if the flux J is known at some concentration difference $(C_1 - C_2)$ of a particular permeant across the membrane. Although reliable methods have been devised to determine membrane thickness (l), only the product $D\beta$ can be derived, giving no independent information regarding the mobility or frictional resistance of solutes within the membrane. Where charged solutes are considered we need to introduce a further term into equation 4.1 to provide for the electrical potential $(E_{\mathrm{I}} - E_{\mathrm{m}})$ across the membrane, where E_{I} is the potential on side I of the membrane from which the flow takes place and E_{m} is the maximum potential within the membrane. Combining the electrical potential with the concentration gradient, the flux J from side I of the membrane to side II is given by

$$J_{\mathrm{I}\rightarrow\mathrm{II}} = PC_{\mathrm{I}}\exp\left[\frac{nF(E_{\mathrm{I}} - E_{\mathrm{m}})}{RT}\right] \tag{4.2}$$

where P is the permeability coefficient and C_{I} the concentration of charged solute on side I of the membrane, n is the number of positive charges on the solute, F is the Faraday constant, R is the gas constant and T the absolute temperature.

Kedem and Katchalsky[2] have analysed the components of the diffusional flux of solutes across cell membranes when this occurs simultaneously with solvent transfer. They showed that solute–solvent interactions and the permeability of the membrane to solvent (in this case water) must be considered in addition to the interplay of forces between solute molecules and the membrane. These factors, together with the molecular size of permeant molecules, are particularly important when solutes diffuse through water-filled channels extending through the membrane. This so-called pore theory of solute transport was strongly supported by Pappenheimer *et al.*[3] (see also Solomon[4]) who showed that solute transport across membranes is a composite of net diffusion and convective flow, both of which are impeded by steric hindrance at the entrance of the pores (which they believed to be located between cells lining the capillary walls rather than across the membrane) and by frictional forces within the pore. If the diameter of such pores is several orders of magnitude greater than the solute or solvent molecules, then transfer of molecules across the membrane occurs by a process referred to as bulk transport or viscous flow. The rate of transfer in this case is directly related to the hydrostatic (or osmotic) pressure applied across the membrane, and the contribution of diffusional flux to total solute transport is of only minor importance. These conditions are only satisfied with pores of large diameter because, as the pores become smaller, bulk transport decreases at a relatively faster rate than diffusional flux. If the ratio of diffusional flux to total flux is compared with pore diameter it is found that the bulk-transport component varies according to about the fourth power of the pore radius and is therefore related to Poiseuille (viscous) flow, whereas the diffusional component is a function of area and varies according to the second power of the pore radius. In practice, transport of most small solutes is restricted to diffusion when the pores have an equivalent diameter of less than 900 pm; when they are larger than this, transport occurs predominantly by bulk flow.

4.1.1 Permeability of Membranes to Water and Gases

Because most lipid bilayer dispersions act as perfect osmometers, measurements of bulk water permeativity can be derived from the rate of swelling or shrinking when an osmotic gradient is applied across the membrane. Bittman and Blau[5] measured the initial volume changes in multilamellar liposomes after establishing concentration gradients of potassium chloride across the bilayers. They showed that the initial rate of permeability to water increased with the introduction of increasing numbers of unsaturated bonds into the constituent phospholipid molecules, indicating that the rate of diffusion is a function of the fluidity of the hydrocarbon chains. The action of cholesterol on water permeativity was predictable on the basis of its effect on mobility of the hydrocarbon chains of phospholipid molecules (see section 2.2.5). Thus the permeability of liposomes consisting of dipalmitoylphosphatidylcholine, dicetylphosphoric acid and cholesterol (64, 4, 32 mole per cent) to water was greater than those containing only phospholipid (96, 4 mole per cent respectively) at a temperature of 38°C (phase-transition temperature = 41°C). There is a marked increase in the permeability of phospholipid vesicles to water when measured above the phase-transition temperature (45°C), but when cholesterol was incorporated into the bilayer water permeativity decreased on heating the dispersion from 38°C to 45°C.

Early attempts to distinguish between bulk transport and diffusional flux of water through membranes were frustrated because of the presence of a boundary layer close to the membrane surface, which exerted an increasing influence on water flux as the permeability of the membrane to water increases. In situations where water permeativity is low, ($P \leqslant 10\ \mu\text{m s}^{-1}$), however, this effect can be safely ignored and accurate rates of transfer by both methods can be determined. Finkelstein and Cass[6] suggested that under these conditions the filtration or osmotic permeability coefficient, P_f closely approaches the relationship

$$\Phi_w = P_f A \phi \Delta C_s \tag{4.3}$$

where Φ_w is the flux of water (in moles/unit time) across a membrane of area A with an osmotic coefficient ϕ and a difference in concentration, ΔC_s, of an impermeant species on either side of the membrane. Similarly the diffusion permeability coefficient, P_d, could be derived from measurements of the flux of isotopically labelled water across the membrane under isotonic conditions, as described earlier in the experiments of Huang and Thompson (section 2.3) using the expression

$$\Phi_w^* = -P_d A \Delta C_w^* \tag{4.4}$$

Φ_w^* in this case represents the flux of isotopically labelled water and ΔC_w^* is the concentration difference of labelled water molecules across the membrane.

Estimates of P_f and P_d across the human erythrocyte membrane, which is highly permeable to water, have been reported by Solomon[4] and Sha'afi *et al.*[7]. To obtain an estimate of osmotic flow they measured changes in red-cell volume by light-scattering in a device that enabled them to record these changes between 100 and 400 ms after establishing an osmotic gradient across the membrane. Subject to the reservations discussed above, they found an osmotic permeability coefficient of 127 μm s^{-1}. A diffusion permeability coefficient had been obtained previously from measurements of the flux of tritiated water across the red-cell membrane, for which they derived a value of 53 μm s^{-1}. The rate of bulk transport is therefore 2.4 times the rate of diffusion, suggesting the presence of channels through the membrane. They then derived an estimate of the equivalent radius of membrane pores, assuming these were of uniform cross-sectional area and that solvent flow proceeded according to Poiseuilles' equation after applying a correction for the restricted diffusion across small diameter pores from the expression

$$r = -a + \sqrt{[2a^2 + (P_f/P_d - 1)\, 8\eta D\bar{V}/RT]} \tag{4.5}$$

where r = pore radius (in Å)

a = radius of water molecule (in Å),
P_f/P_d = ratio of bulk flow to diffusional flow,
η = viscosity of water,
D = diffusion of water in water,
$\bar{V}$ = partial molar volume of water,
R = gas constant,
T = absolute temperature.

By substituting appropriate values of P_f/P_d into equation 4.5 they obtained an equivalent pore radius of 350 pm for channels extending through the human erythrocyte membrane.

Rapid exchange of gases such as oxygen and carbon dioxide is required to sustain the aerobic metabolism of all cells. Some indication of the mechanism by which gases permeate cell membranes has been derived from measurements of gas exchange across detergent films. This system has been exploited by Princen *et al.*[8] to measure the diffusion rates of a number of gases including oxygen and carbon dioxide, and their data were consistent with transport according to the principles of Fick's first law. This conclusion was based on the fact that they could not establish a relationship between the rate of diffusion and the diameter of the gas molecules as predicted by the energy-barrier theory of permeation. Furthermore, large molecules such as CO_2 (diameter 454 pm) diffuse across these membranes at about the same rate as smaller gases like O_2 (diameter 358 pm), suggesting that the transport pathway involves water-filled channels across the membrane film.

The exchange of gases across epithelial tissue of the lung provides a striking example of the rate and dynamics of this process in biological systems. DuBois[9] measured changes in the composition of alveolar gas in the human lung during one breathing cycle and these are illustrated in figure 4.1. It can be seen

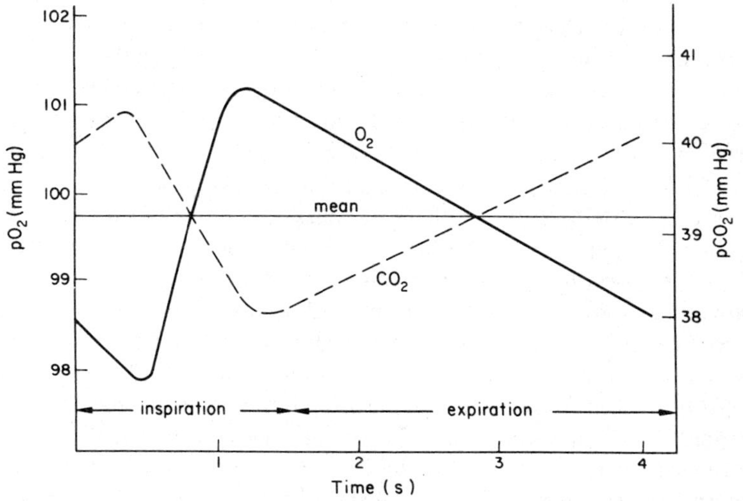

Figure 4.1 Partial pressure of oxygen and carbon dioxide in human alveolar gas during one respiratory cycle (data from reference 9).

that during the first part of inspiration, the alveolar concentration of CO_2 rises to a maximum and O_2 drops to a minimum as dead-space air enters the alveoli. The partial pressure of CO_2 then falls and that of O_2 rises as fresh air reaches the chamber, and they achieve minimum and maximum values, respectively, at the commencement of expiration. During this phase the partial pressure of CO_2 rises and the partial pressure of O_2 falls as gas exchange across the epithelial membranes continues while lung volume is decreasing. The fluctuation in alveolar partial gas pressures tends to be modulated by a buffering effect of air in other parts of the lung and by variations in blood flow during the respiratory cycle. The rate at which blood flows through the lung reaches a maximum during inspiration and a minimum during expiration and, consequently, more gas will be available for exchange when lung volume is increasing than when it is decreasing.

4.1.2 Permeativity of Organic Nonelectrolytes

Gas molecules are small (diffusion is proportional to relative molecular mass for small molecules), and, in contrast to most water-soluble compounds, they do not interact strongly with solvents and thus diffuse readily through lipophilic membranes. It appears that one of the major barriers to diffusion of uncharged solutes across cell membranes is the need to break hydrophilic bonds formed in the aqueous solvent in order to pass into the hydrophobic interior of the membrane. In a classic series of experiments by Collander and Bärlund[10] the rate of diffusion of some thirty organic nonelectrolytes across the plasma membrane of the plant cell, *Chara ceratophylla*, were found to be correlated with the olive oil–water partition coefficients. Olive oil was chosen because the properties of the oil were thought to resemble most closely the hydrocarbon region of biological membranes. However, comparisons between partition coefficients of various nonelectrolytes into dimyristoylphosphatidylcholine bilayers and different organic solvents, including olive oil, suggest that this solvent is too hydrophobic and lower alcohols like octanol appear to be more appropriate[11]. It was also found that partition coefficients of nonelectrolytes depend on the physical structure of the lipid, since solubility in the oil phase is favoured when this is in a liquid-crystalline state. A similar effect has already been noted for electron spin resonance and fluorescence probes, which usually partition to a greater extent into the water phase below the phase-transition temperature of lipid in aqueous dispersions (see section 2.2.3). The permeativity data for *Chara ceratophylla* were subsequently refined by Stein[12] (figure 4.2), who showed that the rate of diffusion (plotted on the ordinate as log permeability coefficient in mm s^{-1} × the square root of the relative molecular mass of the solute) is inversely proportional to the putative number of hydrogen bonds that must be broken for the solute to partition into a hydrophobic environment. The permeability values shown in this figure have been corrected to take account of the enhancing affect of any methylene groups of the permeant solutes. The broken lines indicate the maximum spread of the values and, with the notable

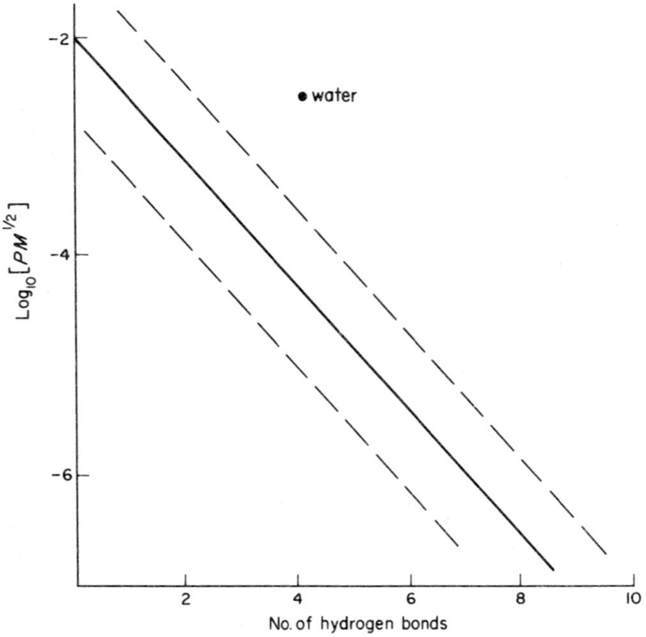

Figure 4.2 Rate of diffusion of solutes across the membrane of *Chara ceratophylla* plotted against the putative number of hydrogen bonds per solute molecule (adapted from reference 12).

exception of water, all these fall within 0.7 log units of the mean regression. A negative correlation indicates that all hydrogen bonds formed between solute and solvent molecules must be broken for transport across the membrane rather than, for example, severing the same number of bonds irrespective of the solute or, alternatively, a different proportion of bonds in each solute. Comparison of the permeativity of organic nonelectrolytes across other cell membranes, such as the erythrocyte, shows a similar trend but there appear to be a number of exceptions, many of which can be accounted for by the presence of specific transporting systems in the membrane.

Lipid dispersions have a high surface area-to-volume ratio and have proved a useful model for studies of the passive permeativity characteristics of both electrolytes and nonelectrolytes across lipid bilayers[13]. Many of these studies employ isotopically labelled permeants because permeability can be measured accurately with small amounts of solute. The experimental technique consists of preparing bilayer vesicles or multibilayer lipid dispersions in a solution containing radioisotopically labelled permeant and removing the untrapped solute by either dialysis against isotonic medium containing non-radioactive solute or molecular sieving through gel columns. When flux rates are low the rate of diffusion can be calculated from the time-course of release of trapped isotope from the lipid structures into a larger volume of dialysis medium, or by using electrodes in the case of electrolytes that are sensitive to particular cations. A more convenient method for solutes with high permeativity is to observe the rate of change of light scattering by the particles as they swell or shrink with changes in osmotic pressure. This technique is feasible becasue the permeability of lipid bilayers to water is eight or nine times greater than for solutes like sugars and charged ions. Moreover, changes in liposome volume appear to be directly proportional to the first power of the reciprocal changes in optical absorbance (for those wavelengths at which absorbance is due entirely to scattered light), a condition that must apply for dependable measurements of the kinetics of liposome swelling and shrinking. Using this method, Cohen and Bangham[14] measured the permeativities of some twenty nonelectrolytes across bilayers formed from mixed dispersions of phosphatidylcholine, phosphatidic acid and cholesterol. A plot of the logarithm of the quotient of permeativity relative to urea and the olive oil/water partition coefficient (a measure of their relative diffusion coefficients) against relative molecular mass produced a highly correlated regression line (figure 4.3).

The dependence of diffusion on relative molecular mass of the solute closely agrees with results obtained with *Chara ceratophylla*[15]. In summarising the available data on the permeativity of water and non-electrolytes across erythrocyte membranes, Sha'afi and his colleagues[16,17] concluded that the transition from aqueous to non-aqueous phase is the major rate-limiting step in the permeativity of a given homologous series of permeants. Permeativity of isomers of individual members of these series, however, depended to a large extent on diffusion rate in the hydrophobic phase. They also suggested that the permeativity of small hydrophilic solutes is limited more by molecular size than the number of potential hydrogen-bonding groups, whereas lipophilic solutes are more dependent on their oil–water partition coefficient which, in turn, is a function of hydrogen-bonding ability and molecular size.

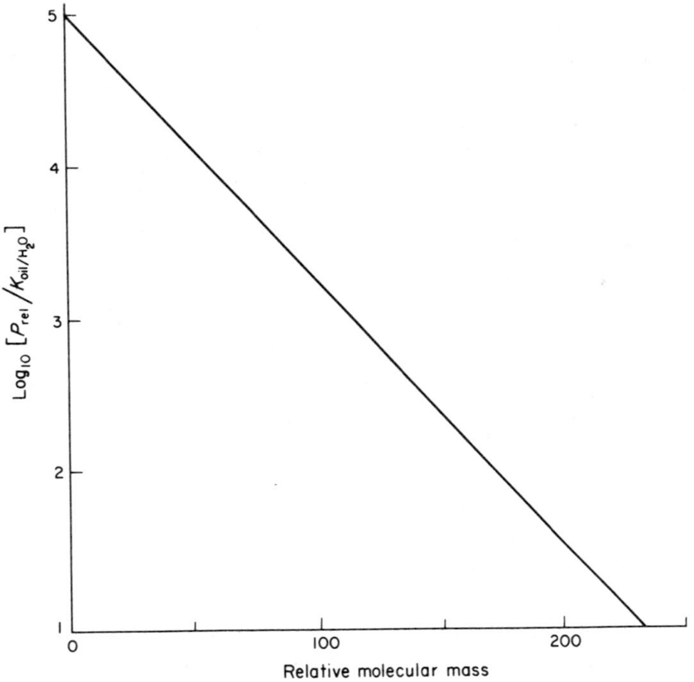

Figure 4.3 The relative diffusion coefficient of solutes across lipid bilayers, plotted as the logarithm of the quotient of the permeativity of a solute relative to urea (P_{rel}) and the olive oil–water partition coefficient as a function of relative molecular mass of solute (from reference 14).

4.1.3 Diffusion of Electrolytes through Lipid Bilayer Membranes

From the above discussion it would seem that the permeability characteristics of lipid bilayers and biological membranes, especially with regard to lipophilic solutes and water, share a number of common features. There are, however, marked differences in the permeativity of hydrophilic solutes, and particularly of inorganic cations, since phospholipid bilayers are generally less permeable to these compounds than are cell membranes. It could be inferred from this that a significant proportion of the flux of hydrophilic solutes is carried through nonhydrophobic pathways in cell membranes and that certain proteins may be directly concerned in this process.

Because inorganic ions and other charged molecules are highly hydrated, the energy required for their release from an aqueous environment and transfer to the hydrophobic region of the membrane provides a major barrier to transport via hydrophobic pathways. Inorganic ions, for example, may be associated with up to twenty water molecules and it has been calculated that the free energy of interaction (charge–dipole) with each water molecule is about 170 kJ mol^{-1}. Corresponding free energies of interaction for hydrogen-bonded water (dipole–dipole interaction) are, for comparison, only about 9 kJ mol^{-1}. Given that ionised groups are surrounded by a tightly bound hydration shell, which probably participates in additional hydrogen-bonding with other water molecules, it seems unlikely that sufficient energy is available to consider permeativity of the nonhydrated ion. The presence of an excess anionic charge on the surface of the bilayer nevertheless appears to assist the permeation of cations across phospholipid bilayers and the effect of these charges on water structure at the bilayer surface cannot be overlooked. Bangham *et al.*[18], for example, showed that the rate of diffusion of sodium or potassium across bilayers of phosphatidylcholine could be very greatly increased by the inclusion of about 5 mole per cent of an amphipathic anion, but diffusion was completely prevented by cationic amphipaths. A plot of the rate of release of potassium from phosphatidylcholine vesicles, containing different amounts of anionic and cationic amphipaths to alter the zeta potential (see 2.2.6), is shown in figure 4.4. Similar permeativities were found for all monovalent cations tested (Li^+, Na^+, K^+, Rb^+, choline$^+$), but permeativities of anions, while appreciably higher than those of cations, did show some differences. Thus, the exchange diffusion rate across phosphatidylcholine bilayers containing 5 mole per cent dicetylphosphoric acid were in the order

$$Cl^- \approx I^- > F^- - > NO_3^- \approx SO_4^{2-} > HPO_4^{2-}$$

and their rates were not affected greatly by surface charge on the bilayer.

The temperature dependence of permeativity, and therefore energies of activation, are also greater for cations than for anions. Bangham *et al.*[18] obtained a value of 63 kJ mole^{-1} for potassium permeativity through phosphatidylcholine:dicetylphosphoric acid bilayers containing 5 mole per cent acidic phospholipid compared to free aqueous diffusion of potassium of about 19 kJ mole^{-1}. Papahadjopoulos and Watkins[19] obtained similar activation energies (63 to 71 kJ mole^{-1}) for potassium permeativity through bilayers consisting of different acidic phospholipids as well as through phosphatidyl-

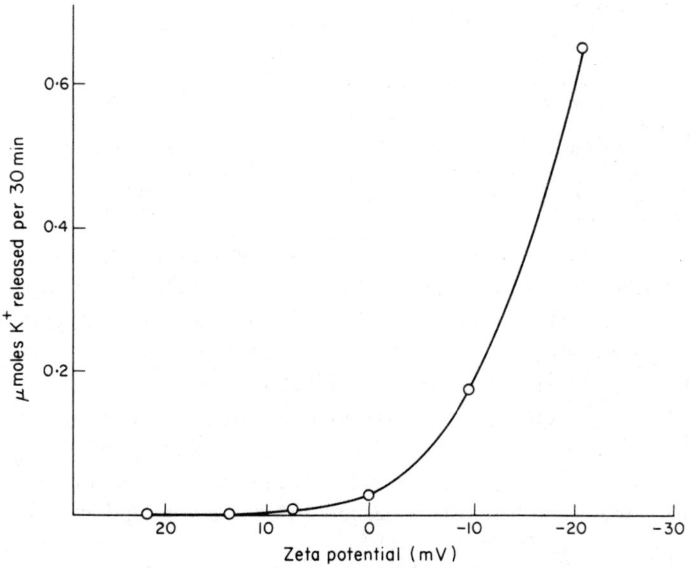

Figure 4.4 The efflux of potassium ions from phosphatidylcholine dispersions containing different proportions of cationic or anionic amphipaths to alter the surface charge. Zwitterionic phosphatidylcholine dispersions have 0 zeta potential (from reference 18).

choline vesicles where the zeta potential had been made positive or negative by including 10 mole per cent stearylamine or stearic acid, respectively. Activation energies for chloride ions, on the other hand, were appreciably lower (21 to 34 kJ mole^{-1}), and not much greater than the energy of activation of chloride ion in free solution.

Papahadjopoulos *et al.*[20] also examined the temperature dependence of sodium diffusion through dipalmitoylphosphatidylglycerol bilayers, particularly in the temperature region where the hydrocarbon chains undergo a phase transition from crystalline to liquid-crystalline. Their results, presented in figure 4.5 show that, at temperatures below the phase-transition temperature, the vesicles were almost impermeable to sodium ions but the diffusion of sodium increased with increasing temperature, reaching a maximum at a temperature coinciding with the mid-point of the lipid phase transition. This was detected by changes in the fluorescence polarisation spectrum of perylene probe added to the lipid dispersion. The diffusion rate thereupon decreased until the phase transition was complete, and then increased again with temperatures exceeding about 45°C.

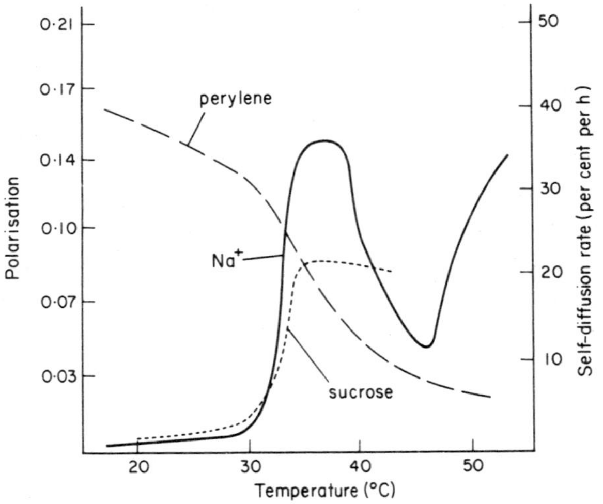

Figure 4.5 The passive diffusion of sucrose and sodium ions from bilayer vesicles of dipalmitoylphosphatidylglycerol at different temperatures. The lipid phase-transition temperature is about 35°C according to fluorescence depolarisation of the perylene probe (data from reference 20).

The permeability to sucrose also increased markedly about the phase-transition temperature, but reached a relatively constant value rather than decreasing with increasing temperature. They explained these observations by suggesting that boundary regions separating domains of phospholipid molecules in crystalline and liquid-crystalline configurations possess a high permeability to sodium ions but not to sucrose. If cholesterol was incorporated into the phospholipid bilayers (in a ratio of 50 mole per cent) the boundary effects of sodium permeability were abolished. The presence of cholesterol in phospholipid bilayers was previously shown to reduce the permeability to other monovalent cations and anions as well as to glycols, glucose and other sugars. The effect of selectively removing cholesterol from erythrocyte membranes upon permeability of the cells to compounds believed to penetrate via a lipophilic pathway is also consistent with these observations. Thus Grunze and Deuticke[21] found that permeability of mammalian erythrocyte membranes to nonelectrolytes such as erythritol and glycerol as well as acetate and propionate ions was unaltered by removing up to 30 per cent of the total membrane cholesterol. Further removal of cholesterol (up to 55 per cent) caused a marked increase in the permeability of the membrane to these solutes. This suggests that hydrocarbon chain mobility may be an important parameter in the passive diffusion of hydrophilic solutes across lipid bilayers.

4.1.4 Ion-conducting Channels in Membranes

The addition of a number of basic proteins to negatively charged phospholipid vesicles is known to increase the rate at which trapped sodium ions diffuse across the bilayer[22,23]. The ability of proteins to increase permeability appears to be related to the relative electrostatic affinity between the protein and the charged phospholipid molecules, which is a property known to enhance the penetration of protein molecules between phospholipid molecules of the bilayer. Much more potent and often specific effects of proteins on permeativity of ions and other hydrophilic solutes have been reported. In biological membranes, certain proteins are believed to be responsible for conducting ions across membranes of excitable cells to create action potentials, while another class of proteins constitutes a group of potent antibiotics, which perturb ion gradients and uncouple oxidative phosphorylation in mitochondria. Studies of these so-called ionophores provide the only available models of the particular chemical and physical interactions that one might expect to find in biological ion transport and membrane excitability. The usual method of identifying these proteins and characterising their properties is to isolate them from biological membranes or other cellular products and observe their effects when incorporated into artificial bilayer membranes. In general, two categories of proteins can be recognised. The first group form permanent channels extending completely across the bilayer and the second constitute transient pathways and act as carriers to shuttle ions from one surface of the membrane to the other. The two mechanisms can usually be distinguished by the fact that conductance by carrier pathways varies proportionately with the concentration of ionophores in the membrane and, when potential gradients across the membrane are low, conductance depends on the rate of diffusion of the complex through the hydrocarbon region of the membrane rather than on the magnitude of the potential gradient. Conductance through channels, on the other hand, is invariably related to the applied potential gradient.

Mueller and Rudin[24] have described early attempts to explore the nature of permeability changes associated with action potentials. They reconstituted an excitable phospholipid bilayer membrane by adding a small amount of excitability-inducing material (E.I.M.), a protein widely distributed in different species and tissues and liberated in appreciable amounts from certain bacteria. The experimental system they used is shown in figure 4.6a. This consisted essentially of two calomel half-cells connected to a recording device and dipping into the two aqueous compartments separated by a lipid bilayer. A constant-current stimulus of voltage V_I could be applied to the same electrodes through resistors in series, R_I, maintained at a value about ten times greater than the

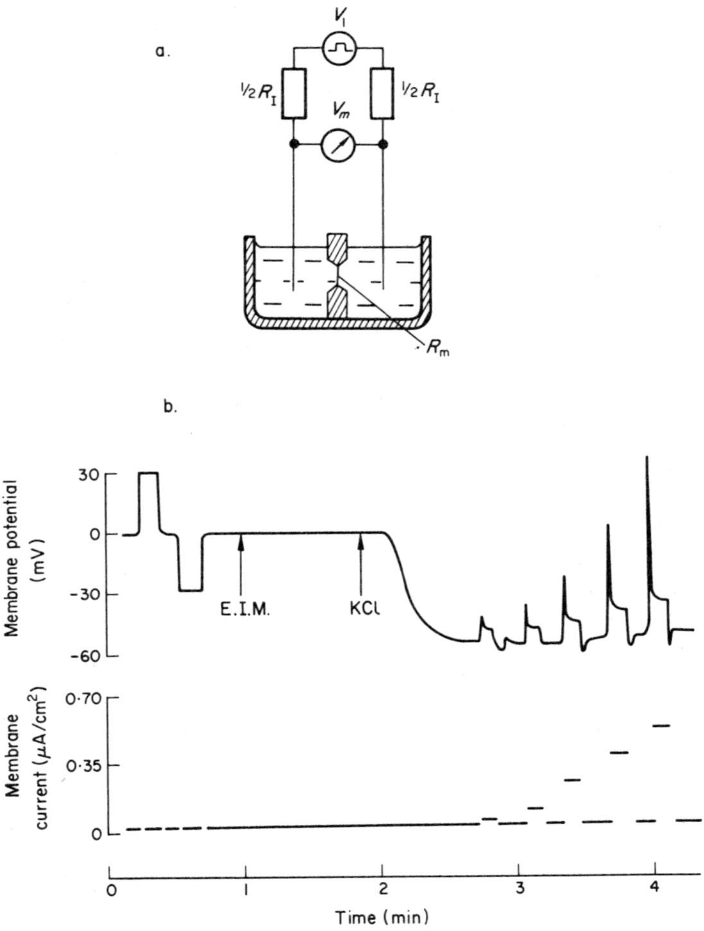

Figure 4.6
(a) Diagram of a device for measuring electrical properties of artificial bilayer membranes.
(b) Changes in membrane potential and membrane current following stimulation with varying voltages, before and after adding excitability-inducing material (E.I.M.) and KCl to one side of the membrane. For details refer to the text (adapted from reference 24).

resistance of the membrane, R_m. When the membrane potential, V_m, is measured, membrane resistance can be calculated from the relationship

$$R_m = \frac{V_m R_I}{V_I - V_m} \tag{4.6}$$

Figure 4.6b shows the development of conductance changes in the bilayer on adding E.I.M. The bilayer in this experiment consisted of a mixture of bovine heart lipids, tocopherol and tetradecane formed across an aperture of area 1 mm^2. Membrane resistance, R_m, was determined (left-hand side of the trace) by supplying a stimulus, V_I, of 54 mV through two $10^8 \Omega$ resistors R_I. The response was to increase membrane potential, V_m, to 30 mV. Under these conditions no change in membrane current (lower trace) is observed, indicating that the membrane is virtually impermeable to ions. The value of membrane resistance $R_m = (0.030 \times 2 \times 10^8/0.054 - 0.030) \times 10^6 = 2.5 \times 10^{14}\ \Omega\ m^{-2}$ When the membrane had again reached its equilibrium potential, small amounts of E.I.M. and KCl (to give a final concentration of 50 mM) were added to one compartment. This produced a decrease of membrane resistance to $1.3 \times 10^{12}\ \Omega\ m^{-2}$, bringing it close to the range normally found in biological membranes (10^9 to $10^{12}\Omega\ m^{-2}$), and membrane potential changed to anew value of 60 mV negative on the side to which the protein and potassium chloride were added. The sign of this potential indicates that the bilayer is more permeable to the cation than to the anion and creates a situation similar to that in resting cells where membrane potential and resistance are high but net membrane current is zero. If depolarising currents of increasing intensity are applied across the membrane, at points indicated by the spikes, the membrane potential approaches voltages corresponding to values of R_m obtained in the absence of a potassium chloride concentration gradient. The membrane potential, instead of maintaining a constant value as it does in the absence of E.I.M. (refer to the initial stimulus recorded on the left), decays at a relatively slow rate consistent with a decrease in membrane resistance. This effect can be explained by an increase in membrane conductance (see lower trace) which serves to diminish the voltage drop across the membrane. Mueller and Rudin suggested that a possible explanation of the increased conductance was that E.I.M. forms a cation-selective channel across the bilayer. In further experiments they were able to demonstrate that a combination of protamine with E.I.M. added to a bilayer containing sphingomyelin formed an anion-conducting channel and action potential changes could be elicited from such membranes.

The presence of channel-forming proteins in gastric mucosal membranes has recently been reported by Sachs *et al.*[25] using an artificial bilayer technique. Proteins were extracted from mucosal membranes with detergents and then incorporated into bilayer membranes. The presence of ionophores among the proteins that exhibited channel rather than membrane carrier characteristics could be demonstrated. Proteins were resolved from the extract by polyacrylamide gel electrophoresis and different fractions were tested individually for ion selectivity in the bilayer system. A fraction of low relative molecular mass was shown to possess anion selectivity with a marked preference for Cl^- over SO_4^{2-} and a fraction of higher relative molecular mass was found to increase the conductance of cations across the membrane. Some other fractions were found to increase conductance across bilayers but showed no discrimination between anions and cations; most other membrane proteins did not cause conductance changes, suggesting that only certain proteins are concerned with ion translocation in membranes of the gastric mucosa.

Hazelbauer and Changeux[26] have succeeded in reconstituting the more complex sodium channel from excitable membranes, which is activated by acetylcholine and its congeners. They treated membranes rich in cholinergic receptor sites, from the electric organ of *Torpedo marmorata*, with detergent to solubilise all membrane components. The detergent (and much of the phospholipid) was then removed by dialysis after which the suspension was unable to trap sodium ions unless supplemented with lipids extracted from native mem-

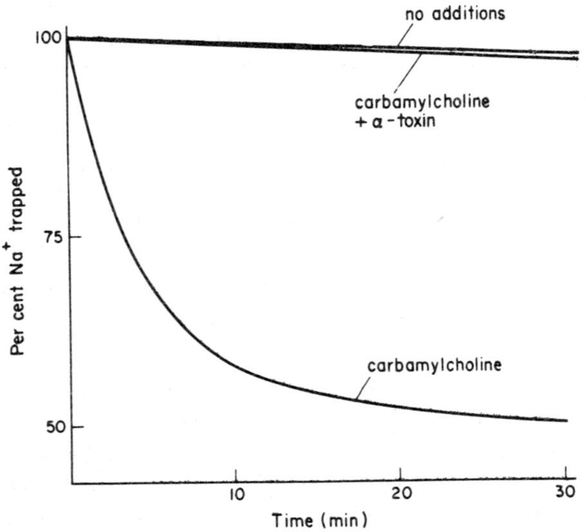

Figure 4.7 Reconstitution of the sodium channel from the electric organ of *Torpedo marmorata.* Release of trapped sodium from reconstituted vesicles by carbamylcholine and its prevention by *Naja nigricollis* α-toxin (adapted from reference 26).

branes. The excitability of the resulting phospholipid enriched vesicles was demonstrated by the ability of the agonist, carbamylcholine, to release sodium ions selectively from the vesicle. The effect of carbamylcholine appeared to be specific because sodium release was completely blocked by the antagonist, *Naja nigricollis* α-toxin (see figure 4.7). Since responsiveness to carbamylcholine requires a recognition step, in this case a binding of the agonist to a receptor site, which then mediates sodium permeativity through the channel, it appears that the regulatory coupling mechanism between the two processes remains intact even after treatment with detergents. Furthermore, the receptor site and channel must be aligned in a functional manner in the reconstituted vesicles, with at least half the receptor sites orientated on the outside surface of the bilayer.

Another group of well-characterised channel-forming compounds includes a series of cyclic peptides, isolated from various bacteria and fungi, which facilitate the diffusion of cations through cell membranes. These compounds can be divided into two groups depending on the structure and, ultimately, the ion selectivity of the channel formed. The less discriminatory group of channel formers include the polyenes, amphotericin and nystatin, consisting of lactone rings with conjugated double bonds. Their action is limited to membranes containing cholesterol, and an assembly of up to ten molecules of the polyene complexed to cholesterol, forming a pore of 500 pm diameter, has been proposed as the likely structure of the channel[27]. The relatively large pore size of these channels permits the passage of fully hydrated ions across the membrane and probably explains why they lack discrimination, in some cases, even between cations and anions.

A more selective group of channel-forming substances includes the gramicidins and N-formyl (ala-ala-gly)$_4$-OMe. These are linear peptides thought to exist in a special kind of helical structure, referred to as β-helix because the hydrogen-bonding pattern between turns of the helix resembles that between chains of a parallel β-pleated sheet structure. The peptides are arranged in the bilayer with hydrophobic amino acid side chains oriented towards the periphery of the molecule and enclosing a central polar core. The pore diameter is too small to permit the passage of hydrated ions and ion selectivity is believed to arise from the presence of polar oxygen groups lining the cores of the helix, which compete with the cations for water molecules bound to the hydrated ion.

4.1.5 Characteristics of Ionic Channels in Nerve Membranes

When nerves or other excitable tissues are stimulated, for example by applying a short pulse of current, the plasma membrane becomes depolarised and, provided the stimulus is greater than a certain threshold value, the change in voltage across the membrane increases to develop an action potential. This change in membrane potentential occurs extremely rapidly but is only a transient phenomenon since the resting potential is quickly restored. Changes in membrane potential in the course of a typical action potential are shown in figure 4.8. Our present view of the mechanism causing these rapid changes in membrane potential stems largely from experiments using a voltage-clamp, a procedure developed many years ago by Cole[28]. Unlike the usual neurophysiological procedure, illustrated in figure 4.8, where potential changes across the membrane are recorded while the nerve is stimulated by a brief depolarising current, the voltage-clamp essentially reverses this process, since a constant potential is established across the membrane and the flow of current is determined. In the unclamped situation, part of the current charges membrane capacitance while the remainder is due to the movement of ions across the membrane. The voltage-

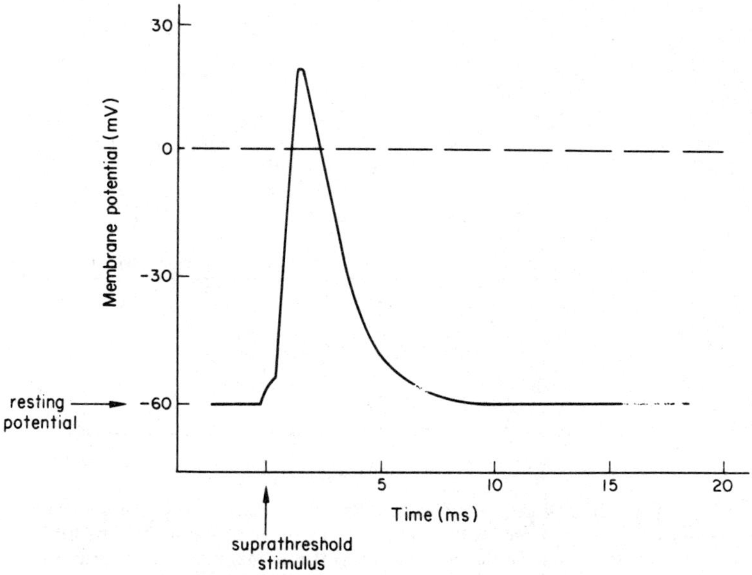

Figure 4.8 Typical changes in membrane potential during the production of an action potential in excitable membranes.

clamp, however, charges membrane capacity to a constant value and the current is carried exclusively by ions moving across the membrane. In practice the passage of as few as 10^6 ions over a period of less than 1 ms can be measured accurately with suitable recording equipment. The particular ions carrying the current can be identified simply by replacing one of the major ions (Na^+, K^+ or Cl^-) with either an impermeant ion or a neutral sugar to maintain osmolarity. Membrane permeability is apparently unperturbed by appropriate ion substitutions and the movement of the remaining ions is largely unaffected. Furthermore, because the net movement of a particular ion will be in a direction of decreasing electrochemical potential, a knowledge of the direction of the gradient under given experimental conditions can be used to predict whether net movement is into or out of the cell. It follows that no membrane current is carried by an ion when the voltage applied across the membrane is at the equilibrium potential of that ion.

Hodgkin *et al.*[29] recorded changes in the ionic current when a squid giant axon was depolarised with a voltage just sufficient to reduce the transmembrane potential to zero (60 mV). Non-myelinated nerves of crustaceans and squid are favoured for this type of experiment because of their large size (1 mm diameter) and apparent similarity to nerves from higher organisms. Typical results of such

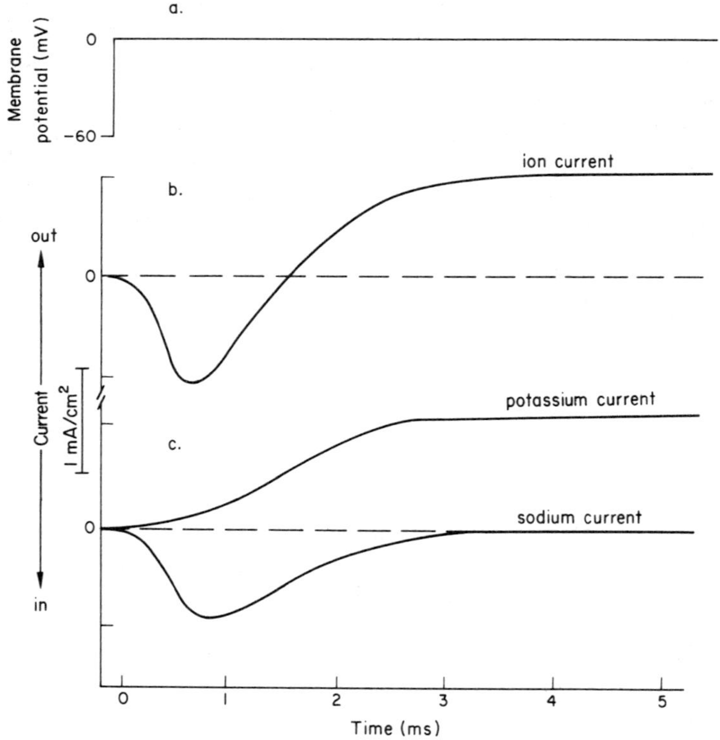

Figure 4.9 A typical dissection of membrane currents across a nerve axon under voltage-clamp conditions. When a voltage clamp is applied to a nerve to reduce the membrane potential to zero (a) the resulting ion current across the axonal membrane varies in a time-dependent manner (b). Using appropriate ion substitutions, the ion current can be shown to consist of two major components: one due to the movement of potassium ions out of the axon and the other due to the movement of sodium ions into the axon (c).

an experiment are illustrated in figure 4.9 where it can be seen that, during the first few milliseconds after applying the voltage-clamp (figure 4.9a), the current first flows inwards and then is reversed in favour of a substantial and continuous outward current (figure 4.9b). On substitution of 90 per cent of the sodium ions in the bathing solution with choline ions (an amount that equalises sodium activity on either side of the axonal membrane), Hodgkin and Huxley[30] were able to resolve this current into two major components, one carried by sodium ions and the other by potassium ions (figure 4.9c). A further minor component was recognised, which they termed the leakage current, but they were unable to ascertain the precise source of this current. Hodgkin and Huxley[31] subsequently demonstrated that the relationship between voltage and current at any point in time is linear, so that an expression similar to Ohm's law could be formulated to equate ionic movements with the driving forces. The current carried by each particular ion, potassium for example (I_K), is given by

$$I_K = g_K(V - V_K) \tag{4.7}$$

where g_K is the potassium conductance, V is the membrane voltage or potential and V_K is the equilibrium voltage, at which the net movement of potassium is zero. Likewise, the sodium current, I_{Na} and the leakage current, I_L can be expressed in similar terms. The total ionic current, I_{ionic}, is obtained by summing each of the individual currents so that

$$I_{ionic} = I_K + I_{Na} + I_L \tag{4.8}$$

The conductances, g_K and g_{Na} depend on voltage but the *rate* of activation of each ion is a function of the degree of depolarisation of the membrane. The leak conductance, on the other hand, is independent of voltage and remains constant with time. The time-course of the sodium conductance following voltage-clamp according to Hodgkin[32] is the product of two independent processes, an activation factor known as m^3 and an inactivation factor referred to as h. These factors can be considered as gates, both of which must be open for sodium ions to pass through. The mathematical solution to the sodium conductance is therefore

$$g_{Na} = \bar{g}_{Na} m^3 h \tag{4.9}$$

where $\bar{g}_{Na}$ is the maximum sodium conductance across the membrane, m^3 is the proportion of all sodium transport sites with open activation gates and h, those with open inactivation gates. The expression recognises the existence of a fixed number of sodium (and potassium) transport sites in the membrane, more of which are recruited or activated as the potential is increased until the potential approaches the depolarising voltage when $g_{Na} = \bar{g}_{Na}$ and all sites are fully conducting. Referring to the sodium conductance in figure 4.9c, the time-course of events can be described operationally as a rapid activation when the m^3 and h gates are open and sodium moves through the channel and into the nerve. At the same time the h gate begins to close (sodium inactivation), eventually obstructing the channel and preventing the passage of sodium ions. Potassium permeativity only exhibits activation under voltage-clamp conditions, and hence the existence of only one gate need be postulated. This is known as the n^4 activation factor. Potassium conductance, g_K is given by

$$g_K = \bar{g}_K n^4 \tag{4.10}$$

where g_K is the maximum potassium conductance and n^4 is the proportion of sites available to conduct potassium ions. The time-course of potassium activation suggests that the gate operates more slowly than sodium activation, a feature that can be observed by comparing curves for sodium and potassium currents in figure 4.9c.

More recent studies have shown that complete removal of sodium from the external medium does not always block action potentials particularly, for example, in mollusc nerve-cell bodies. Standen[33] has analysed the ion currents of snail neurones by voltage-clamp techniques and showed that the early inward current can be carried by calcium ions as well as, or instead of, sodium ions. The two ions appear to use the same channel but the possibility of a parallel transporting mechanism with identical kinetics could not be disproved. The kinetics of calcium influx into heart muscle, however, does seem to be different from that of sodium, since the inflow is sustained beyond sodium inactivation and causes a pronounced plateau in the action potential of cardiac cells. Since action potentials can be generated in the absence of calcium, it is thought that sodium may provide an effective substitute[34]. An equally complicated picture is beginning to emerge from detailed studies of the potassium current. Part of the potassium current, for example, becomes inactivated in many preparations with time constants of one second or more and, while not of concern in isolated action potentials, the process may be necessary to sustain long-term firing characteristics of the cell. Using potassium-sensitive microelectrodes to measure the outward flux of potassium from snail neurones, Lux and Eckert[35] concluded that part of the apparent potassium inactivation was due to a slow inward current presumably carried by either sodium or calcium ions.

Apart from the fact that permeativity changes of sodium and potassium in nerve axons are kinetically independent, there is other evidence indicating that the transport site for each ion is different. Hille[36] has recently collated this evidence and concluded that there are discrete membrane channels for sodium and potassium ions, and separate channels which conduct the leakage current. The discovery that certain fish toxins such as tetrodotoxin and compounds like tetraethylammonium salts selectively block sodium and potassium conductance, respectively, has provided the most convincing demonstration so far. Indeed these agents have proved versatile tools in establishing the dimensions and chemistry of the individual channels. The permeativity of cations through the sodium channel, for example, has been measured in the frog sciatic nerve (myelinated) under voltage-clamp conditions when the nerve is bathed in a sodium-free medium containing the test ion and with tetraethylammonium ion blocking about 95 per cent of the potassium channels. The permeativity ratio is then determined when sodium is present after appropriate corrections have been applied to account for leakage currents (see table 4.1). Likewise, permeativity of other cations through the potassium channels can be estimated when bathing solutions contain high concentrations of the test ion and tetrodotoxin (more than

90 per cent of sodium channels may be blocked by tetrodotoxin). From the nature of the ions passing through the respective channels it has been calculated that the cross-sectional dimensions of the sodium channel are about 300 × 500 pm, while the entrance to the potassium channel is about 800 pm but is constricted at one point to a diameter of only 300 pm. This means that sodium ions passing through the sodium channel can retain three or four bound water molecules, whereas potassium must relinquish all but two molecules of hydration when passing through the potassium channel. It may also be noted from table 4.1 that sodium permeativity through the potassium channel and potassium permeativity through the sodium channel are both restricted.

Table 4.1 Permeativity ratios of alkali metal ions and ammonium ions through sodium and potassium channels of frog nerve.

Cation (X)	Sodium channel P_X/P_{Na^+}	Potassium channel P_X/P_{K^+}	Diameter of ion (pm)
Li^+	1.110	<0.018	120
Na^+	1.000	<0.010	190
K^+	0.086	1.000	266
Rb^+	<0.013	0.910	296
Cs^+	<0.012	<0.077	338
NH_4^+	0.016	0.130	300

data from Hille 37, 38

Additional information about the character of these channels has been derived from the effects of pH on ion conductance. Bathing frog nerves in solutions of low pH, for example, causes a marked decrease in sodium and potassium permeativity in a voltage-dependent manner[38,39]. The titration curve of sodium permeativity gives a pK_a of 5.4 for a single dissociable group and, taken together with the voltage-dependent characteristics, suggests that a carboxyl group is situated part-way through the membrane rather than at the entrance to the channel. The pK_a of the titratable group(s) associated with the potassium channel, however, is somewhat lower (pK_a 4.4) and may involve more than one ionisable group. The most likely molecular arrangement within the potassium channel is thought to be a ring of oxygen atoms with a diameter of 300 pm producing a region of low electrostatic field strength.

Protease digestion of axonal membranes has been employed as a probe of the channel gating mechanism, and the process most sensitive to this procedure appears to be sodium inactivation. Armstrong *et al.*[40] perfused axons internally with pronase and found that sodium inactivation could be completely destroyed, whereas sodium and potassium conductances were unaffected by pronase digestion. One of their experiments, illustrated in figure 4.10, shows the effect of pronase treatment on the I_{Na} and I_K plotted as a function of membrane potential. A squid giant axon was clamped for 12 ms with the required voltage and I_{Na} was obtained from the peak early current (before potassium activation) and I_K from the steady-state current towards the end of the voltage-clamp (after sodium inactivation). Curve A of figure 4.10 shows the relationship between

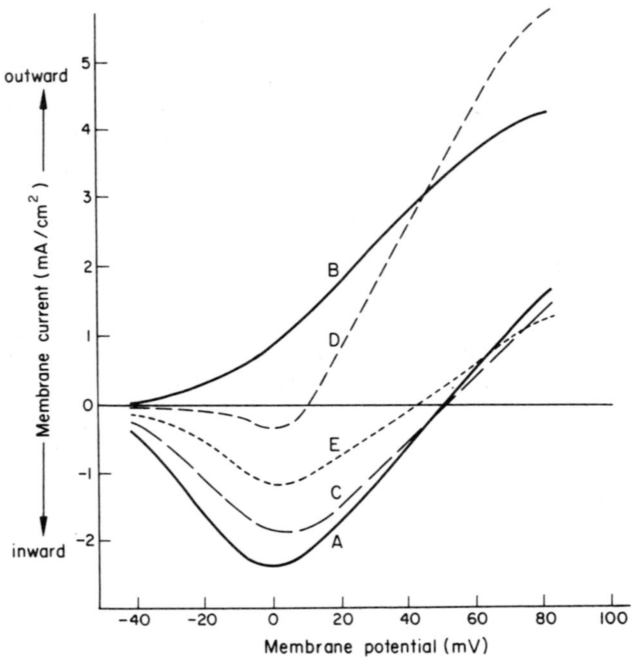

Figure 4.10 The sodium (A) and potassium (B) currents in squid giant axons as a function of membrane potential (at 9°C). The equivalent sodium (early)current (C) and potassium (late) current (D) recorded in a nerve 8 min after internal perfusion with pronase. Curve E represents the amount of the total steady-state current conducted by sodium ions after pronase digestion (D − B). (adapted from reference 40).

I_{Na} and membrane potential and curve B represents I_K. The inward sodium current is maximum when the membrane potential is zero and the equilibrium potential (V_{Na} from the sodium equivalent of equation 4.7) is 54 mV; these parameters did not change when the measurements were repeated after an eight-minute internal perfusion of the axon with pronase (curve C). The slight reduction in current carried by sodium ions could be accounted for by the complete destruction of some sodium channels but the shape of the curve suggests that sodium activation is unaffected by internal pronase perfusion. The steady-state current (curve D), however, is altered remarkably by pronase treatment. The ionic current is still directed inward when the membrane potential is negative, but, as this becomes more positive, the current increases markedly, cutting curve B at a point close to the equilibrium potential for sodium conductance. When the steady-state current after pronase treatment is subtracted from the steady-state current before digestion (D − B), curve E is obtained, which represents I_{Na} or the fraction (in this experiment 60 per cent) of channels that no longer exhibit sodium inactivation and continue to conduct sodium ions under steady-state conditions. It was concluded from these results that the *h* gate consists of a protein located at the inner entrance of the sodium channel and that the *m* and *n* gates are either situated in regions not accessible to the protease or not susceptible to pronase attack.

4.2 Facilitated Diffusion

The observed rates of permeativity of certain solutes across cell membranes are considerably greater than might be predicted from their chemical and physical properties; that is, they are either too large to penetrate through membrane channels or too hydrated to permeate readily through a lipophilic barrier. Glucose and glycerol, for example, move across the human erythrocyte membrane at a rate several orders of magnitude faster than would be expected from either the permeability characteristics of the membrane or the relative molecular mass and putative number of hydrogen-bonding groups of each solute. This suggests that diffusion of such molecules through the membrane is facilitated in some way.

Facilitated diffusion can be distinguished from free diffusion across membranes because finite rates of transport can be achieved if enough permeant is added to one or other side of the membrane. That is, the diffusion process can be saturated and Fick's first law of diffusion as expressed in equation 4.1 can no longer be applied. This property indicates that molecules do not diffuse through specialised, highly permeable regions of the membrane but are restricted to passage through a fixed number of specific transporting sites (note, however, the boundary effects on permeativity of cations through phospholipid vesicles discussed in section 4.1.3). Facilitated diffusion also differs from active transport because no metabolic energy is required to drive the process although it could be argued that energy is expended in maintaining an intact membrane. The net transport of molecules is therefore always in a direction of decreasing electrochemical potential. Indeed, this gradient provides the driving force for facilitated diffusion irrespective of whether this is a gradient of the solute transported or another molecular species to which movement of the first is coupled.

A number of factors point to the participation of membrane-associated carrier proteins in facilitated diffusion. Firstly, a number of enzyme poisons are known to markedly and specifically inhibit facilitated diffusion when added in concentrations low enough to exclude any nonspecific effects on membrane permeability. These reagents effectively decrease the number of transport sites without disturbing the rate of transport through unreacted sites, a situation analogous to irreversible inhibition in enzyme kinetics. Secondly, the transport sites possess a remarkable chemical specificity and in many cases can even discriminate between optical isomers of certain solutes. It is probable, although not conclusive, that a protein would be required to achieve the necessary degree of selectivity. Nevertheless, some permeants appear to be transported at the same site and typical competitive-type inhibition can usually be demonstrated. D-glucose and D-galactose, for example, both compete for the same transport site in the erythrocyte membrane but neither of the L-isomers of these sugars is transported at appreciable rates across the membrane[41]. The molecular specificity exemplified by this sugar-transporting system provides additional evidence against simple diffusion since the D- and L-isomers would be expected to diffuse at approximately the same rate.

4.2.1 Kinetics of Facilitated Diffusion

Because of the similarity between the kinetics of facilitated-diffusion systems and enzyme reactions the flux J of a permeant in one direction can be formulated along classical Michaelis–Menten lines to give the expression

$$J = \frac{SV_{max}}{K_m + S} \tag{4.11}$$

where S is the concentration of permeant molecules on the side of the membrane from which the flow takes place, V_{max} is the maximum rate of flux and K_m is the permeant concentration when $J = V_{max}/2$. Similarly the action of non-competitive inhibitors of transport is like that of irreversible enzyme inhibitors, since both tend to decrease V_{max} by blocking a proportion of the active sites. Competitive inhibitors of transport cause an increase in K_m because of the apparent decrease in affinity of permeant molecules for the transport site without affecting the maximum rate of flux that can be achieved. The flux expression given in equation 4.11 of course, only refers to the movement of solute in one direction, and, because of the symmetry of the transport process, solutes may move with equal facility in the reverse direction. The net flux of solute is given by

$$\text{net flux} = J_{\text{I}\to\text{II}} - J_{\text{II}\to\text{I}} \tag{4.12}$$

where I and II refer to opposite sides of the membrane.

A comparison of the rates of unidirectional flux (equation 4.11) and net flux (equation 4.12) of solutes through membranes is helpful in establishing the existence of facilitated diffusion systems. Glucose transport across the human erythrocyte membrane has been investigated by LeFevre and McGinniss[42], who measured unidirectional and net fluxes using radioisotope techniques. In these

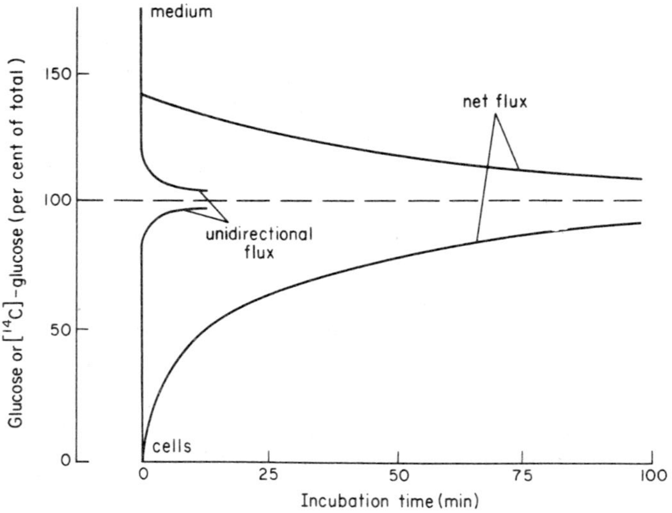

Figure 4.11 The rate of uptake of glucose (net flux) and [^{14}C]-glucose (unidirectional flux) into human erythrocytes. Refer to text for experimental conditions (taken from reference 42).

experiments (see figure 4.11), glucose-free erythrocytes were incubated at 20°C in a solution of glucose and samples were removed at intervals to measure sugar concentration in the cells and suspending medium. Glucose transport was inhibited with mercuric chloride immediately on withdrawing samples so that a more efficient separation of cells could be achieved without altering the glucose concentration of either compartment. A parallel incubation was performed at 37°C to establish an equilibrium concentration of glucose on either side of the erythrocyte membrane. This suspension was subsequently adjusted to 20°C and a small amount of radioisotopically labelled glucose was added to the medium and samples taken again, this time to determine the distribution of radioisotopically labelled glucose between the cells and the medium. The rate of uptake of unlabelled glucose in the first part of the experiment gives a measure of the net flux of sugar across the membrane, whereas the tracer experiment measures unidirectional flux. The rationale for the measurement of unidirectional flux with labelled glucose is that radioactive glucose taken up from the external medium is exchanged for unlabelled molecules moving in the opposite direction, since the tracer is diluted in a large intracellular pool of unlabelled glucose. It can be seen from figure 4.11 that the time for half-equilibration of radioactive glucose is some fifty to a hundred times faster than the corresponding half-equilibration time for unlabelled molecules entering the cells. Clearly, if glucose permeativity were by simple diffusion the rate of entry of glucose into the cell, measured as the net transfer of unlabelled molecules across the membrane, should be the same as that measured by the unidirectional flux of labelled glucose. Given that glucose has a high affinity for a membrane-transport site the results can only be explained by the operation of a facilitated diffusion process for glucose across the erythrocyte membrane.

Another type of facilitated diffusion, known as counter transport, has been demonstrated in some tissues and the feature of this process is that sugar appears to be transported against a concentration gradient although no expenditure of metabolic energy is involved. The phenomenon was first described by Park *et al.*[43] and typically, when cells are equilibrated with a nonmetabolisable sugar such as 3-O-methylglucose and then a high concentration of glucose is added to the extracellular medium, the glucose analogue is transported out of the cell against a concentration gradient. The energy for this transport is provided by the coupled diffusion of glucose into the cell moving in a direction of decreasing concentration.

4.2.2 Mechanism of Facilitated Diffusion

So far we have considered that diffusion takes place at a fixed number of membrane-transport sites. These sites are generally believed to be carrier molecules, which are mobile within the membrane and shuttle solutes back and forth between opposite sides of the membrane. When a gradient of a particular solute exists, the carrier picks up molecules from the surface of higher concentration and deposits them on the other side, returning as free carrier to collect more solute. The mobile carrier hypothesis is supported by inhibitor studies which have shown that the rate of transport by facilitated diffusion is reduced if the membrane surface to which net transport takes place is exposed to nonpenetrating inhibitor. No cases have been reported where diffusion in one direction is affected by nonpenetrating inhibitors to a greater extent than movement in the opposite direction.

One might anticipate that movement in one direction would proceed independently of movement in the opposite direction; however, comparison of the diffusion rates of some solutes suggests that this may not always be the case. Glucose diffusion through the erythrocyte membrane is a good example. In this process the rate of diffusion of carrier in the membrane appears to depend on whether or not solute molecules are complexed to the carrier. This was the conclusion of Levine *et al.*[44] in experiments on glucose transport across the human erythrocyte membrane. Cells were preloaded by incubation in media containing 2.6 mM radioisotopically labelled glucose and the efflux of labelled glucose was measured after transfer to media containing different concentrations of unlabelled glucose. Efflux was determined at 0°C and samples taken for counting after thirty seconds incubation because of the high rate of glucose flux across the membrane. As the concentration of extracellular glucose increased from 0 to 15 mM the unidirectional flux of labelled glucose from the cells almost doubles and achieves a constant rate when the unlabelled glucose concentration of the medium is greater than 15 mM. This effect was explained as a faster rate of diffusion of the complexed carrier compared to the glucose-free carrier. Thus more carrier would be loaded in the influx direction as the external glucose concentration is raised, and would therefore be available to return with labelled molecules picked up from inside the cell.

4.2.3 Mitochondrial Uncoupling Agents. Proton Carriers

Mitchell[45] was the first to propose that agents that uncouple oxidation reactions in mitochondira from phosphorylation of nucleotide were lipid-soluble weak acids that translocate protons across the inner mitochondrial membrane, thereby collapsing the proton-motive force connecting the two processes (see section 5.3.6). Uncoupling anions, or negatively charged complexes of the anion, are believed to act as carriers by associating with protons from the intermembrane space, transporting them across the inner mitochondrial membrane and releasing them to the matrix.

Complexes commonly take the form HA_2^-, where two anion molecules (A^-) conduct a single proton across the membrane, but HAB^- types, where A^- and B^- represent different anions, have also been reported. Experimentally, HA carriers can be distinguished from complex forms by observing changes in conductivity of artificial bilayer membranes with increasing concentrations of uncoupling agent. LeBlanc[46] has used this method to examine the carrier conformation of the uncoupling agent, carbonylcyanide *m*-chlorophenylhydrazone. He found that conductivity of phosphatidylcholine:cholesterol:decane bilayers increased proportionately with increasing uncoupler concentration over the range, 0.1 to 10 μM and the highest conductivities were recorded when the pH was higher than the pK_a of the anion. This result may be expected from proton carriers in the HA form. Other uncoupling agents, consisting mainly of substituted benzimidazoles, produce changes in bilayer conductivity according to the square of the uncoupler concentration, and transmembrane conductivity is maximum when the pH is equal to the pK_a of the acid. The highest conductance at pH = pK_a is predicted for complexes of the HA_2^- form[47]. The uncoupling agents, 2,4-dinitrophenol and 5,6-dichloro-2-trifluoromethylbenzimidazole were investigated by Foster and McLaughlin[48] and both were found to be of the HA_2^- form. Moreover, when both uncoupling agents were present, a carrier complex of the HAB^- type was formed by association of the two acids in the bilayer membrane.

There have been conflicting views on the comparative effectiveness of uncoupling agents in mitochondria and their effects on proton translocation across phospholipid bilayers. Bakker *et al.*[49] measured the concentration of six different uncoupling agents required to produce half-maximal uncoupling of rat liver mitochondria and then tested their ability to alter fluxes across bilayer dispersions and increase conductance across artificial bilayer membranes. Two liposome systems were investigated. The first involved the transfer of protons from ascorbic acid on the outside of the liposome to reduce ferricyanide trapped inside the dispersion in the presence of an electron carrier. The second was to observe the rate of osmotic swelling when liposomes were placed in an isosmotic potassium acetate solution in the presence of the potassium ionophore, valinomycin. Both liposome systems produced excellent agreement with the mitochondrial uncoupling activity of these agents. Bilayer conductivity measurements, however, showed no correlation with effectiveness in mitochondrial systems. This may be partially explained by the anomalous behaviour of the HA_2^--type carriers in artificial bilayer systems, and it is possible that agreement will be obtained when precautions are taken to ensure that appropriate pH and other conditions of measurement are adopted[48].

4.2.4 Ionophores as Membrane Carriers

The ability of certain peptides of microbial origin to create ion-conducting channels across artificial and biological membranes has already been mentioned (see section 4.1.4). There is, however, another group of compounds from similar sources, which facilitates the diffusion of ions across membranes, but instead of forming membrane channels they operate as ion carriers by moving ions from one membrane surface to the other. All these compounds have antibiotic activity and they include valinomycin, nigericin, alamethicin, enniatins A, B, C and the actin homologues. The common structural feature of these compounds is that they all possess a cyclic conformation, with hydrophobic residues orientated towards the periphery of the molecule, when placed in an apolar environment. They partition in apolar solvents to form lipid-soluble ion complexes in which ions are held inside the ring by ion–dipole interactions. Because of the relatively fixed dimensions of the hydrophilic region they tend to discriminate between different ions, at least more so than many of the channel-forming antibiotics.

The ion-transport specificity of cyclic peptide antibiotics in mitochondrial and erythrocyte membranes and phospholipid bilayer vesicles has been evaluated by Henderson *et al.*[50] and compared with the transport properties of the polyene

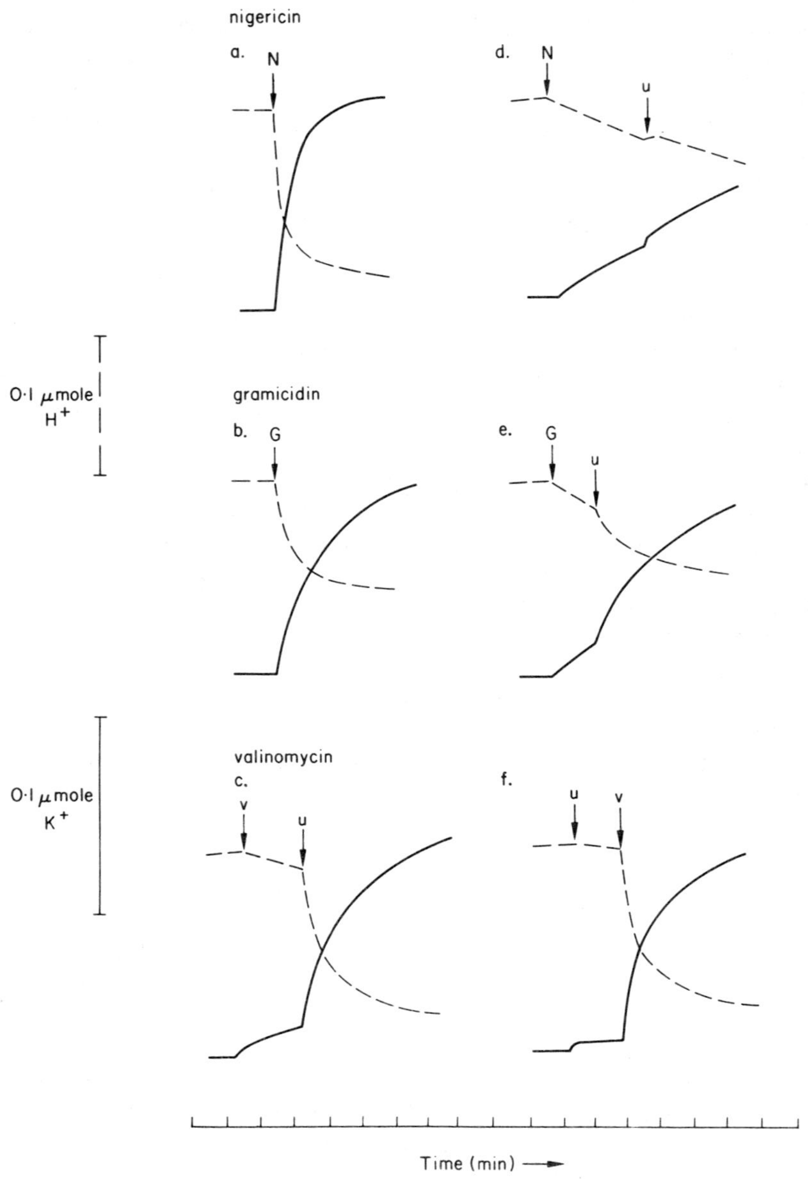

Figure 4.12 Facilitated diffusion of potassium ions from phospholipid bilayer liposomes by ionophores. The concentration of nigericin, N, and gramicidin, G, in (d) and (e) is 1/100 that in (a) and (b), respectively. The addition of uncoupling agent, u, carbonyl cyanide *p*-trifluoromethoxyphenylhydrazone is indicated by arrows (adapted from reference 50).

antibiotics. The changes in ion permeability produced in each of the three membrane systems were qualitatively similar, suggesting that the drugs operate independently of other proteins in biological membranes. The net flux of trapped potassium from liposomes consisting of phosphatidylcholine plus 10 mole per cent dicetylphosphoric acid, and the corresponding influx of protons when different classes of ionophore are present, is shown in figure 4.12. Non-permeant counter ions (mucate and choline) were present on either side of the bilayer, so potassium cannot diffuse out of the liposome unless protons can enter in exchange. Each of the ionophores shown in figure 4.12 differ in the degree to which they can facilitate K^+-H^+ exchange across the bilayer. Nigericin and, to a lesser extent, gramicidin D facilitate the efflux of potassium and influx of protons but valinomycin is almost completely ineffective unless a proton carrier, carbonylcyanide *p*-trifluoromethoxyphenylhydrazone, is added as well. It may be noted parenthetically that uncoupling agent alone does not catalyse K^+-H^+ exchange (figure 4.12f). Further evidence that nigericin lacks discrimination between protons and potassium ions can be seen in figure 4.12d because, even when added in low concentration to liposomes, the exchange of ions across the bilayer is not enhanced by uncoupling agent. Gramicidin D, on the other hand, is seen to show a preference for potassium because K^+-H^+ exchange is increased by uncoupling agent (figure 4.12e), and in tracer experiments $^{42}K^+$-K^+ exchange was rapid when the uptake of protons into the liposomes was slow. The same fast tracer exchange was observed in the presence of valinomycin but this ionophore showed no affinity for protons. Similar studies of proton exchange for other monovalent cations have been performed, and the cation selectivities of the three ionophores are presented in table 4.2.

Table 4.2 Cation and proton selectivity of three antibiotic ionophores

Drug	Cation selectivity	H^+ permeability
nigericin	$K^+ \approx Rb^+ > Na^+ > Cs^+ > Li^+$	induces free H^+ permeation
gramicidin D	K^+, Rb^+, Cs^+, $NH_4{}^+$, Na^+, Li	limited H^+ permeation
valinomycin	K^+, Rb^+, $Cs^+ > NH_4{}^+$, Na^+, Li	none

data from reference 50

Not all antibiotic ionophores are restricted to forming carrier complexes with monovalent cations; beauvericin for example, appears to associate preferentially with calcium ions. Prince *et al.*[51] have compared the cation specificity of this antibiotic with valinomycin and ennaitin A in a liposome preparation and in a naturally occurring bacterial chromophore system. Liposomes containing trapped bacteriochlorophyll and the fluorescent compound 9-aminoacridine were prepared at pH 6 and the external pH was increased to establish a proton gradient across the bilayer. As in the experiments outlined above, the addition of mitochondrial uncoupling agents does not disturb the proton gradient because translocation is opposed by an electrochemical potential gradient. Protons can be released, however, if they are exchanged for an equivalent number of cations, in this case at a rate dependent on the facilitated diffusion of the antibiotic ion-carrier complex. The rate of collapse of the pH gradient in the presence of different cations and ionophores was monitored by the rate of fluorescence quenching of the trapped dye. The other system used to explore cation specificity consisted of intact bacterial chromophores. Carotenoids present in these membrane-enclosed structures undergo a shift in spectral absorbance on exposure to light and this is associated with the generation of an electric potential across the membrane which is positive with respect to the inside of the organelle. Prince and his colleagues[51] exploited this effect to reverse the process, that is to use ion gradients to drive a spectral shift of careteniods in darkened chromophores. Their results confirmed previous observations that neither valinomycin nor ennaitin A could translocate divalent cations across membranes. Beauvericin, in contrast, showed a marked preference for calcium ions (but not barium, strontium, magnesium, aluminium or lanthanide ions) in both membrane systems. The transport specificity in order of preference was $Ca^{2+} > K^+ > Cs^+ > Li^+ > Na^+$. Interestingly they found that calcium was transported as a singly charged species rather than as the fully charged ion.

4.3 Active Transport

In much the same way that simple diffusion cannot account for the rates of carrier-mediated diffusion across membranes, the rate of permeation, and particularly the steady-state distribution of some of the most important physiological substances, defy explanation in terms of spontaneous diffusion. The low intracellular sodium and high potassium concentrations found in nearly all cells, and the uptake of a variety of sugars, amino acids and other substrates necessarily involves transport against electrochemical potential gradients. Such transport can only be accomplished at the expense of metabolic energy and an important feature of active-transport systems is that they are all unidirectional in their energy-requiring functions. That is, active transport takes place only in one direction across membranes; movement in a direction of decreasing electrochemical potential is accomplished by some form of diffusion or it can be linked to the movement of other permeants travelling against an electrochemical potential gradient.

Since active transport requires the continuous supply of energy, usually in the form of ATP, procedures that inhibit metabolism effectively block active transport. Transport can often be reactivated in metabolically inert cells with an exogenous supply of ATP, in many cases only when this is present at a particular side of the membrane. The question of how this chemical energy is utilised to drive the transport of solutes has been the subject of great interest and intensive study over the past few years. In general, energy-yielding reactions could facilitate the rate of transfer of a carrier–permeant complex or free carrier across the membrane. Alternatively, these reactions may induce conformational changes in the carrier, which then influence the association and dissociation of permeant and carrier on respective surfaces of the membrane.

There are several ways in which the expenditure of chemical energy may be linked to solute transport. The simplest form is a direct transfer of a single molecular species from one side of the membrane to the other at a rate depending directly on the energy consumed. More complex connections may involve the coupled transport of a number of solutes simultaneously across the membrane, in which an obligatory attachment of two different solutes to the transport site is required before either can be translocated across the membrane. Linked systems have been described in which an electrochemical potential gradient of one solute, established by active transport, provides the energy for the transport of a second solute moving against a gradient. In some cases the coupling appears to be even more remote since an electrochemical potential gradient established for a linked solute may be used to drive the transport of a third permeant across the membrane. Other linked systems involve an exchange process in which the same carrier is responsible for the movement of different permeants in opposite directions across the membrane.

The rate of active transport, irrespective of the particular mechanism involved, can be obtained from the usual Michaelis-Menten formulation (equation 4.11). As with facilitated diffusion, transport takes place at a fixed number of transport sites in the membrane to which permeants have a characteristic affinity. Nevertheless additional complications arise in linked systems, where concentrations of more than one permeant must be considered in any overall kinetic evaluation of the process.

4.3.1 Group Transfer of Sugars in Bacteria

Wild-type bacteria grown on a substrate of glucose contain active-transport systems for glucose, mannose and their analogues but not for a large number of other sugars. If the medium is supplemented with different sugars, including pentoses, hexoses or disaccharides, specific permeases catalysing the active uptake of these sugars may be induced. While several mechanisms have been proposed to explain the uptake of sugars by bacteria[52], the majority of permeases appear to be intimately connected to a cytoplasmic phosphorylating system and are transported by a process referred to as group transfer. Apart from a tight coupling to metabolic activity, the distinguishing feature of the transport process is that phosphorylated derivatives of the sugar are formed during transit through the membrane. It may be noted that sugar phosphates are the active inductants of permease systems and not the corresponding free sugar. The mechanism of phosphorylation and details of the phosphotransferase have been published in a

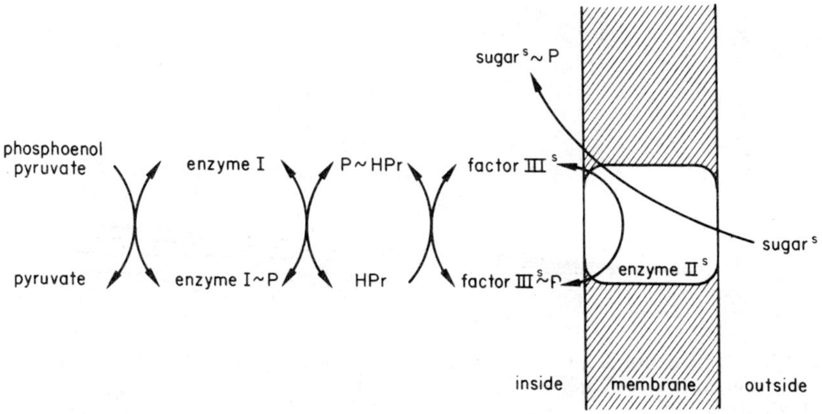

Figure 4.13 The reaction sequence of sugar uptake by group transfer into *Staphylococcus aureus* (gram-positive). Superscript s refers to sugar specific and induceable components.

series of papers by Roseman and his colleagues[53–8]. The phosphorylating system of gram-positive bacteria is shown in figure 4.13; it consists briefly of three cytoplasmic proteins Enzyme I, and two histidine-containing phosphocarrier proteins known as HPr and Factor III. A fourth component, Enzyme II, is firmly bound to the bacterial membrane and relies on other membrane components, notably lipids, for activity. The membrane-bound counterpart of the permease system in gram-negative bacteria is believed to consist of a combined Factor III-type protein (II_B) and an Enzyme (II_A). The constitutive II_A component has been resolved in *Escherichia coli* into three subfractions, one of which is specific for glucose, another for mannose and the third for fructose. The activity of all these components appears to depend on a specific interaction with phosphatidylglycerol in the membrane.

The phosphorylation and translocation of sugars commences with the transfer of phosphate from the glycolytic intermediate, phosphoenolpyruvate, to HPr and the reaction is catalysed by Enzyme I. The phosphate is then transferred from HPr to Factor III, which possesses three possible phosphorylation sites. Finally, phosphorylated Factor III interacts with Enzyme II and sugar to form a ternary complex resulting in the transfer of phosphate to the sugar catalysed by Enzyme II. Enzyme II therefore functions as the solute-recognition site and catalyses the phosphotransfer reaction, while Factor III serves as the energy-coupling component of the system. An interesting aspect of the energetics of the transport process is that the production of sugar phosphates during transport represents a appreciable saving of free energy, particularly since sugar phosphates serve as direct precursors for subsequent metabolic reactions. The standard free energy of hydrolysis of phosphoenolpyruvate under physiological conditions is

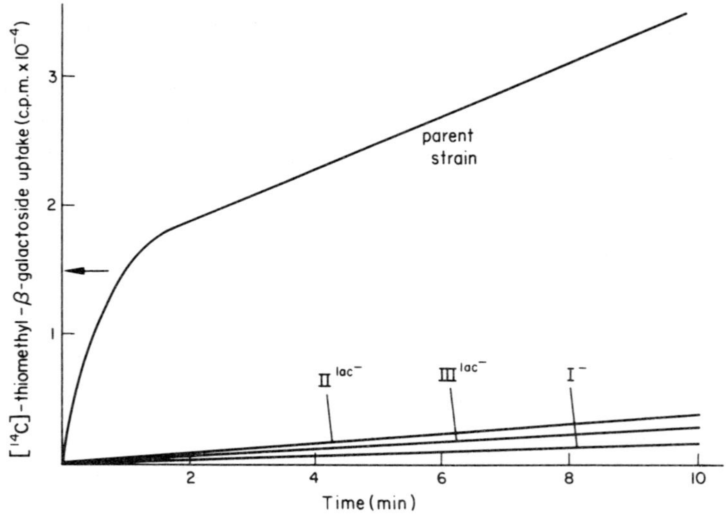

Figure 4.14 The uptake of nonmetabolisable lactose analogue, thiomethyl-β-galactopyranoside into mutant strains of *Staphylococcus aureus* compared to the wild-type parent. The equilibrium concentration of sugar across the membrane is shown by the arrow on the ordinate. The uptake of the non-metabolisable glucose analogue, methyl α-D-glucopyranoside in II^{lac-} and III^{lac-} mutants was identical to the parent strain, but no uptake of this sugar was observed in I^- mutants.

about $-$ 59 kJ mole^{-1} and most of this is likely to be conserved in P—N bonds of phospho-HPr and phosphorylated Factor III (approximately $-$ 50 kJ mole^{-1}). The corresponding free energy of hydrolysis of phosphate attached to carbon-6 of hexoses is about $-$ 15 kJ mole^{-1} and it has been suggested that some of the excess free energy is used to drive the translocation of sugar across the membrane. How this energy transduction is accomplished is not known.

Simoni and Roseman[58] have isolated mutant strains of *Staphylococcus aureus* that are unable to transport lactose, and examined their capacity to transport other sugars in order to determine which components of the permease system were specific to lactose. The uptake of the nonmetabolisable lactose analogue, thiomethyl-β-galactopyranoside, by strains lacking specific components of the phosphotransferase system is shown in figure 4.14 (no strains deficient in HPr have yet been isolated). Deficiency in any one of these components results in a failure to accumulate the sugar, however: when the uptake of the glucose analogue, methyl α-D-glucopyranoside was examined, it was found that transport in II^{lac-} and III^{lac-} mutants was indistinguishable from the parent strain. In constrast, I^{lac-} mutants were unable to transport the glucose analogue, suggesting that Enzyme I is common to the transport pathway of both sugars, whereas Enzyme II and Factor III are sugar-specific and induceable components.

4.3.2 Sugar and Amino Acid Permeases in Animal Cells

Group transfer of sugars across the bacterial cell membrane has no known parallel in animal cells, where the mechanism of sugar transport is fundamentally different. It has been known for many years that transport of various sugars and amino acids across epithelial and other cell membranes requires the presence of sodium ions; substitution of lithium, potassium, choline or other monovalent ions for sodium reduces transport capacity. Since many factors affecting the transport of sugars appear to have the same effect on amino acid transport it could be inferred that both solutes are transported by a similar mechanism[59]. In a kinetic study of the unidirectional flux of radioactive L-alanine across the brush border membrane of epithelial cells in the rabbit ileum, Curran *et al.*[60] concluded that the transport system consisted of a ternary complex between the membrane carrier, alanine and sodium ions in which both solutes are transported simultaneously across the membrane. When choline replaces sodium in the mucosal solution the maximum rate of alanine uptake remains unchanged but the Michaelis constant for the process increases in proportion to the residual sodium concentration. Furthermore, influx of alanine was found to be accompanied by the movement of sodium into the cell, and the amounts of alanine and sodium crossing the membrane approached equality at high sodium concentrations, suggesting that the two fluxes are linked. In subsequent studies it was reported that *p*-chloromercuriphenyl sulphonate, a reagent that reacts with sulphydryl groups, affects the reaction kinetics in much the same way as choline substitution for sodium ions, indicating that sulphydryl residues of the carrier protein are concerned with sodium binding[61].

The mechanism of coupling between sodium and organic solutes during transport is not known. One theory maintains that the energy required to sustain the osmotic work of transporting sugar or amino acid against a concentration gradient is derived entirely from the electrochemical potential gradients of ions across the membrane. No direct chemical energy is needed because one kind of osmotic energy (ion gradient) is simply exchanged for another (solute gradient). An alternative theory considers that chemical energy is more directly coupled to the membrane components responsible for active sugar and amino acid transport. In other words, the membrane transduces chemical energy from metabolic sources into osmotic energy. Kimmich[62] has carefully examined the relationship between sodium and sugar transport in the small intestine and reported that no conclusive evidence was yet available to support either the direct or indirect coupling mechanisms. The fact that active sodium extrusion from epithelial cells takes place at the serosal membrane of the cells, whereas sugar and amino acid transport sites are located in the mucosal membrane, tends to support the ion-gradient hypothesis. Moreover, inhibitors of solute transport such as ouabain appear to act exclusively on the outer surface of the serosal membrane, which is consistent with an effect on the sodium transport system. The effect of inhibitors of sodium transport, as expected, is to cause an accelerated efflux of solute rather than affecting uptake, suggesting that the solute-transporting mechanism is not the primary site of inhibition. Some observations, however, are at variance with an indirect energy coupling process, since it has been noted that solute transport in certain cases can proceed despite a reversal of the sodium gradient across the membrane. It has been argued that a combined sodium–potassium gradient could supply the necessary chemiosmotic energy to drive solute uptake, but even when both sodium and potassium gradients are reversed it has been shown that Ehrlich ascites tumour cells can accumulate α-aminoisobutyrate, a nonmetabolised amino acid, against a concentration gradient[63]. This has led to the suggestion that transport of certain amino acids in these cells may be coupled more directly to the utilisation of ATP[64].

The identification of glucose-transport sites in different membranes has been investigated using specific competitive inhibitors of glucose transport. Among the most potent of these are the phenolic glucosides, such as phlorizin, which compete with D-glucose for binding to the glucose carrier. In the luminal brush border membrane of the renal proximal tubule, two types of phlorizin-binding site can be distinguished, one with high affinity and another with low affinity for the inhibitor[65]. The high-affinity site is most likely to be the glucose transporter because sodium is required for binding and glucose competes with inhibitor for the binding site. The high-affinity site was shown by Silverman[66] to bind sugars in a pyranose ring in the chair conformation with hydroxyl groups on carbon-3 and carbon-6 orientated as in D-glucose. Other sugars with similar conformation such as 2-deoxy-D-glucose and D-galactose share this transporter system. Another sugar-transport site in the brush-border membrane has a preferred substrate of D-mannose. The stereospecific requirements for sugar-binding to the antiluminal surface of renal tubular cells is different from that of the brush-border (luminal) membrane because, in addition to the pyranose-ring configuration, hydroxyl groups must be present on carbon-1 and carbon-2, and if present on carbons-3 or -6 they must be orientated equatorially as they are arranged in D-glucose.

The active transport of amino acids, at least across membranes of intestinal epethelia, falls into three groups depending on affinities for particular classes of amino acids. They include a neutral amino acid permease specific for amino acids with only one carboxyl and one amino group, a pathway for basic amino acids specific for dibasic amino acids (with the exception of cystine) and finally an amino acid–glycine permease for glycine, N-substituted glycine derivatives, proline and hydroxyproline. Another permease for dicarboxylic amino acids has been reported in Ehrlich ascites tumour cells, but so far this permease has not been reported in other tissues.

4.3.3 Calcium Transport across the Sarcoplasmic Reticulum

The sarcoplasmic reticulum is a network of tubules, vesicles and cysternae surrounding myofibrils, primarily of skeletal muscle but also found in other types of muscle. The membranes of these structures are specialised to accumulate and bind calcium ions. Depolarisation of the sarcoplasmic membrane leads to the translocation of sequestered calcium from the sarcoplasmic reticulum to the myofibrils providing the connecting link between excitation and contraction of the muscle[67]. The degree of specialisation is exemplified by the fact that membranes of the sarcoplasmic reticulum contain only a few proteins and between 60 and 80 per cent of the total membrane protein is Ca-activated ATPase. A number of other highly acidic proteins have been isolated from the membrane, the most notable being calsequestrin, which is capable of binding up to forty moles of calcium per mole of protein. Calsequestrin is bound to the external surface of the membrane and is readily released by calcium sequestering agents.

The mechanism of calcium transport has been investigated by examining the disposition of Ca-activated ATPase in the sarcoplasmic membrane using selective proteolysis and lactoperoxidase–iodination techniques[68–70]. In conjunction with ultrastructural studies, it was found that Ca-activated ATPase is an intrinsic membrane protein orientated on the external surface of the membrane and, in negatively stained preparations, appears as a 4 nm diameter sphere attached to the membrane by a stalk 2 nm long. Tryptic digestion of intact vesicles of sarcoplasmic reticulum reduces ATPase activity and cleaves the enzyme complex (relative molecular mass 110 000) into three fragments. Two peptides, of relative molecular mass 55 000 and 60 000, are the initial products and the latter is subsequently cleaved to yield two peptides of relative molecular mass 33 000 and 24 000 respectively. The larger of the two serves as a phosphorylation site in the transport process and both are iodinated by lactoperoxidase added to the outside of intact membrane vesicles. The fragment of relative molecular mass 55 000 is not iodinated from either surface of the membrane nor is it degraded further by trypsin, suggesting that it is located predominantly in the hydrocarbon region of the membrane.

The uptake of calcium by vesicular preparations of sarcoplasmic reticulum requires magnesium and involves hydrolysis of high-energy phosphate bonds; ATP is presumed to be the natural substrate for the reaction. The exact relationship between the energy provided by hydrolysis of the phosphate ester and transport has not been established, mainly because ATPase activity is not tightly coupled to calcium transport in many preparations. A ratio of two moles of calcium transported per mole of ATP hydrolysed is commonly observed but this may be somewhat less than the true ratio because values as high as ten have been obtained in some cases. A possible reaction sequence for translocation of calcium across the sarcoplasmic reticulum has been proposed by Martonosi[71]. Briefly, this involves the interaction of ATP with the enzyme, which becomes phosphorylated in the presence of calcium ions. The reaction is readily reversible, since a fast ATP–ADP exchange is usually observed. The phosphorylation takes place at the external surface of the membrane, after which the phosphorylated enzyme–calcium complex is presumed to undergo a conformational change, resulting in the translocation of calcium to the inner surface of the membrane, where it is released from its binding site. The dissociation of calcium from the transport site is accompanied by hydrolysis of the phosphate bond, and the cycle is completed when the enzyme returns to its original conformation with the active centre orientated on the outer surface of the membrane.

While it is true that we know little of the possible conformational changes associated with phosphorylation and translocation of calcium, both these reactions appear to depend on the nature of the interaction of the enzyme with other components of the membrane, in particular the phospholipids. Membrane vesicles treated with phospholipase C, for example, lose ATPase activity and the ability to transport calcium, but both functions can be restored by adding phosphatidylcholine to the residue. Other phospholipids and detergents can reactivate the ATPase, but phosphatidylcholine seems to be required in the penultimate reaction involving cleavage of the enzyme–phosphate bond. Studies of the effect of temperature on Ca-activated ATPase activity and rates of calcium transport have suggested that the physical properties of the membrane in the vicinity of the enzyme is an important factor in the overall transport process. Inesi *et al.*[72] have reported changes in the slope of Arrhenius plots of the motion of several spin-label probes attached either to the Ca-activated ATPase or to fatty acids interpolated into the membrane. These changes are centred about a temperature of 20°C and the protein-bound probe also undergoes a marked transition about 40°C. The nature of these changes is obscure considering that the liquid-crystal to crystal phase transition in the membrane as a whole would be expected somewhere in the region of −20°C. Nevertheless, these transitions appear to be correlated with ATPase activity and rates of calcium transport. Thus, when the temperature is increased above 20°C, the activation energy for both processes decreases from 116 kJ $mole^{-1}$ to 71 kJ $mole^{-1}$, indicating that changes in the hydrocarbon region of the membrane alter the activity of the enzyme. The transition observed at 40°C in spin-probes attached to the enzyme is associated with an uncoupling of ATPase activity from calcium transport, since ATPase activity and passive calcium flux from the vesicles continues unabated above 40°C, where calcium uptake ceases. It was concluded from the electron spin resonance spectrum that a change in protein conformation is responsible for the uncoupling.

4.3.4 Exchange Transport of Sodium and Potassium

Almost without exception, cells maintain a low intracellular sodium concentration and a high intracellular potassium concentration and, at equilibrium, the distribution of ions across the membrane establishes a net negative electrical potential inside the cell. Since the net movement of the respective ions is in a direction opposing their particular electrochemical potential gradients, transport can be regarded as an active process. Indeed the energy dependence can be readily demonstrated if cellular metabolism is inhibited; sodium rapidly enters the cell under these conditions and potassium leaks out. These movements are passive and the rate therefore depends on the permeability characteristics of the membrane. This means that the distribution of sodium and potassium under steady-state conditions is a function of the activity of the pumping process relative to the leakage of ions along their respective electrochemical potential gradients.

The transport process involves movement of the two ions in opposite directions across the plasma membrane but that need not require both movements to be active. It could be that active transport of one or other ion results in the passive redistribution of the other according to a potential gradient created across the membrane, provided the membrane is sufficiently permeable to allow diffusion of the ion. Despite the existence of a net negative potential inside the cell, indicating that charge is unequally distributed across the membrane, direct chemical evidence implicates both ions in a coupled exchange-transport mechanism. Radiochemical studies of the reaction have shown that the γ-phosphate of ATP is transferred to an acyl linkage on the membrane-bound carrier protein in a reaction requiring the presence of magnesium ions (the substrate in effect is ATP-Mg^{2+}), and is stimulated by sodium ions. The phosphorylated ATPase enzyme is believed to be an intermediate product in the active transport of ions across the membrane. The final step involves hydrolysis of the phosphate ester bond, a reaction dependent on magnesium but stimulated this time by potassium ions. The potassium-dependent phosphatase releases inorganic phosphorus and the transport protein is restored to its original form. Bader *et al.*[73] examined the activity of (Na^+-K^+)-ATPase from a large number of different species and tissues and found that although the specific activity of the enzyme varied considerably the activity of the phosphorylation reaction was nearly always the same as the corresponding potassium-dependent phosphatase activity. The exception appears to be erythrocytes of genetically determined low-potassium (LK) individuals of certain ruminant species in which activity of the potassium-dependent phosphatase, as measured by the hydrolysis of the synthetic substrate *p*-nitrophenyl-phosphate, was considerably lower than the genetically dominant high-potassium (HK) type[74].

The ATPase enzyme is firmly bound to the plasma membrane and never recovered in soluble form from tissue homogenates. Attempts to dislodge it from the membrane invariably reduce enzyme activity. This tends to suggest that other membrane components are required for activation although the question of whether specific interactions are involved has not yet been satisfactorily resolved. Removal of phospholipids from the membrane with detergents, organic solvents or digestion with phospholipases A or C partially or, in some cases, completely inactivates the enzyme. Activity can often be fully restored by adding mixed phospholipid extracts to delipidated preparations of the enzyme and phosphatidylserine, in particular, seems to be the most effective phospholipid. The absolute dependence of enzyme activity on interaction with phosphatidylserine, however, has been challenged recently by de Pont *et al.*[75], who treated a bovine brain preparation possessing high (Na^+-K^+)-ATPase activity with phosphatidylserine decarboxylase and succeeded in converting 99 per cent of phosphatidylserine to phosphatidylethanolamine without loss of enzyme activity.

The precise mechanism of energy transduction, as with most other transport processes, is not clear, but some information can be deduced from the location of binding sites for substrate and ions, which helps to establish the probable orientation of the enzyme in the membrane. Whittam and Ager[76], working with erythrocyte membranes and Baker[77] with crab nerve showed that ATP and sodium ions interact with sites on the enzyme accessible only from the cytoplasmic surface of the membrane whereas potassium interacts with an external site. The presentation of substrate or ions to alternate sides of the membrane does not activate the enzyme. The location of particular binding sites is consistent with the transporting function of the enzyme, that is, extrusion of sodium and accumulation of potassium from the extracellular fluid. It should be noted that ADP and inorganic phosphorus from the potassium-dependent phosphatase reaction are both released from the cytoplasmic surface of the membrane even though the potassium-binding site is located on the external surface of the membrane[78].

The stoichiometry of the reaction in erythrocytes has been measured by determining the fluxes of sodium and potassium across the plasma membrane corresponding to hydrolysis of intracellular ATP. In general, three sodium ions are extruded and two potassium ions are taken up by erythrocytes for each molecule of ATP hydrolysed. In nerve and muscle, ratios of three sodium ions extruded per ATP hydrolysed are usually observed, but the amount of potassium entering these cells often depends on the particular experimental conditions. One consequence of the ion-transport ratios is that multiple ion and substrate interactions must be required during ion translocation. These complexities are probably reflected in the molecular configuration of the enzyme. Conventional methods for estimating relative molecular mass, including gel filtration and sedimentation velocity, have given values ranging from 190 000 to 500 000 for different preparations of the enzyme and even if the minimum relative molecular mass is taken it is likely that the enzyme consists of a number of subunits. If 250 000 relative molecular mass is assumed, with a density of 1.3 for the protein, the molecule in a spherical configuration would have a diameter of 8.5 nm, which would be large enough to penetrate the lipid bilayer completely and expose binding sites simultaneously on both surfaces of the membrane.

4.4 Summary

One of the most important physiological properties of cell membranes is that they are selectively permeable. In general, large proteins are not permeative but smaller molecules are able to diffuse through at a rate determined by their size and solubility in the hydrophobic region of the membrane. Cell membranes are relatively permeable to gases and water mainly because of their small size. The presence of an ionisable group greatly reduces permeativity, since such groups are highly hydrated. Inorganic ions diffuse at a very slow rate through artificial phospholipid bilayer membranes, suggesting that cell membranes are intrinsically resistant to the penetration of ions. Proteins isolated from biological membranes and others derived from certain micro-organisms are believed to form water-filled channels through which selected ions may pass. Highly selective channels for sodium and potassium are present in nerves and other excitable membranes, which permit the rapid and transient passage of these ions across the membrane when stimulated with a brief depolarising voltage. These changes in permeability result in the development of action potentials and constitute the basis of membrane excitability. Voltage-clamp procedures have provided a useful method of analysing the ionic movements through the respective channels and have been used to characterise the associated gating mechanisms.

The penetration of certain solutes across cell membranes is facilitated by the presence of specific transport proteins in the membrane. In contrast to free diffusion, transport can be saturated when sufficient solute is present to fully engage all the transport sites. These proteins combine with solute one side of the membrane and diffuse across to the other side, releasing the solute and returning for more. The net flux is always in a direction of decreasing electrochemical potential. The diffusion of protons across membranes can be facilitated by lipophilic weak acids and this phenomenon has been used to explain the action of these compounds in uncoupling oxidative phosphorylation in mitochondria. Ionophores obtained from microbial sources also act as specific ion carriers across cell membranes.

When metabolic energy is required to transport solutes across membranes the movement is referred to as active transport. The active process is unidirectional in that energy is expended only in moving solutes against an electrochemical potential gradient. Certain inorganic ions, sugars and amino acids are the main substrates transported in an active manner. The uptake of sugars by bacteria occurs by a mechanism known as group transfer in which free sugar is phosphorylated during passage across the membrane. Sugar and amino acid transport in eukaryotes is linked to the influx of sodium ions moving in a direction of decreasing electrochemical potential and this energy is thought to be utilised in driving the transport of substrate against a gradient. Membranes of the sarcoplasmic reticulum are particularly adapted to transport calcium. The transport protein is a calcium-activated ATPase and is phosphorylated by ATP in the presence of calcium, which is then translocated across the membrane and released on the other side when the phosphate bond is hydrolysed. Sodium and potassium ions are transported across the plasma membrane by an exchange-transport process. The ATPase transport enzyme is phosphorylated in the presence of sodium ions at the cytoplasmic surface of the membrane and subsequently dephosphorylated in a reaction dependent on potassium ions presented at the outer surface of the membrane. Usually three sodium ions are extruded and two potassium ions are taken up by the cell for each ATP molecule consumed. Both calcium-activated and sodium–potassium-activated ATPases rely on interactions with membrane phospholipids for their activity.

References

1. E.M. Wright and R.J. Pietras. Routes of nonelectrolyte permeation across epithelial membranes. *J. Membr. Biol.*, **17** (1974), 293–312
2. O. Kedem and A. Katchalsky. A physical interpretation of the phenomenological coefficients of membrane permeability. *J. gen. Physiol.*, **45** (1961), 143–79
3. J.R. Pappenheimer, E.M. Renkin and L.M. Borrero. Filtration, diffusion and molecular sieving through peripheral capillary membranes. A contribution to the pore theory of capillary permeability. *Am. J. Physiol.*, **167** (1951), 13–46
4. A.K. Solomon. Characterization of biological membranes by equivalent pores. *J. gen. Physiol.*, **51** (1968), 335s–64s
5. R. Bittman and L. Blau. The phospholipid–cholesterol interaction. Kinetics of water permeability in liposomes. *Biochemistry,* **11** (1972), 4831–9
6. A. Finkelstein and A. Cass. Permeability and electrical properties of thin lipid membranes. *J. gen. Physiol.*, **52** (1968), 145s–72s
7. R.I. Sha'afi, G.T. Rich, V.W. Sidel, W. Bossert and A.K. Solomon. The effect of the unstirred layer on human red cell water permeability. *J. gen. Physiol.*, **50** (1967), 1377–99
8. H.M. Princen, J.Th.G. Overbeek and S.G. Mason. The permeability of soap films to gases. II. A simple mechanism for monolayer permeability. *J. Colloid Interface Sci.*, **24** (1967), 125–30
9. A.B. Du Bois. Alveolar CO_2 and O_2 during breath holding, expiration and inspiration. *J. appl. Physiol.*, **5** (1952), 1–12
10. R. Collander and H. Bärlund. Permeabilitatis-studien an *Chara ceratophylla. Acta bot. fenn.* **11** (1933), 1–114
11. J.M. Diamond and Y. Katz. Interpretation of nonelectrolyte partition coefficients between dimyristoyl lecithin and water. *J. Membr. Biol.*, **17** (1974), 121–54
12. W.D. Stein, *The Movement of Molecules Across Cell Membranes,* Academic Press, New York and London (1967)
13. A.D. Bangham, M.W. Hill and N.G.A. Miller. Preparation and use of liposomes as models of biological membranes. In: *Methods in Membrane Biology,* vol. 1 (E.D. Korn ed.), Plenum Press, New York and London (1974), pp 1–68
14. B.E. Cohen and A.D. Bangham. Diffusion of small non-electrolytes across liposome membranes. *Nature, Lond.*, **236** (1972), 173–4
15. W.R. Lieb and W.D. Stein. Biological membranes behave as non-porous polymeric sheets with respect to the diffusion of non-electrolytes. *Nature, Lond.*, **224** (1969), 240–3
16. R.I. Sha'afi and C.M. Gary-Bobo. Water and non-electrolytes permeability in mammalian red cell membranes. *Prog. Biophys. molec. Biol.*, **26** (1973), 103–46

17. P. Naccache and R.I. Sha'afi. Patterns of nonelectrolyte permeability in human red blood cell membrane. *J. gen. Physiol.*, **62** (1973), 714–36

18. A.D. Bangham, M.M. Standish and J.C. Watkins. Diffusion of univalent ions across the lamellae of swollen phospholipids. *J. molec. Biol.*, **13** (1965), 238–52

19. D. Papahadjopoulos and J.C. Watkins. Phospholipid model membranes. II. Permeability properties of hydrated liquid crystals. *Biochim. Biophys. Acta,* **135** (1967), 639–52

20. D. Papahadjopoulos, K. Jacobson, S. Nir and T. Isac. Phase transitions in phospholipid vesicles. Fluorescence polarization and permeability measurements concerning the effect of temperature and cholesterol. *Biochim. Biophys. Acta,* **311** (1973), 330–48

21. M. Grunze and B. Deuticke. Changes in membrane permeability due to extensive cholesterol depletion in mammalian erythrocytes. *Biochim. Biophys. Acta,* **356** (1974), 125–30

22. H.K. Kimelberg and D. Papahadjopoulos. Interactions of basic proteins with phospholipid membranes. Binding and changes in the sodium permeability of phosphatidylserine vesicles. *J. biol. Chem.*, **246** (1971), 1142–8

23. H.K. Kimelberg and D. Papahadjopoulos. Phospholipid–protein interactions: membrane permeability correlated with monolayer penetration. *Biochim. Biophys. Acta,* **233** (1971), 805–9

24. P. Mueller and D.O. Rudin. Resting and action potentials in experimental bimolecular lipid membranes. *J. theor. Biol.*, **18** (1968), 222–58

25. G. Sachs, J.G. Spenney, G. Saccomani and M.C. Goodall. Characterization of gastric mucosal membranes. VI. The presence of channel-forming substances. *Biochim. Biophys. Acta,* **332** (1974), 233–47

26. G.L. Hazelbauer and J-P. Changeux. Reconstitution of a chemically excitable membrane. *Proc. natn. Acad. Sci. U.S.A.*, **71** (1974), 1479–83

27. T.E. Andreoli. The structure and function of amphotericin B-cholesterol pores in lipid bilayer membranes. *Annal. N.Y. Acad. Sci.*, **235** (1974), 448–68

28. K.S. Cole. Dynamic electrical characteristics of the squid axon membrane. *Archs. Sci. physiol.*, **3** (1949), 253–8

29. A.L. Hodgkin, A.F. Huxley and B. Katz. Ionic currents underlying activity in the giant axon of the squid. *Archs. Sci. physiol.*, **3** (1949), 129–50

30. A.L. Hodgkin and A.F. Huxley. The dual effect of membrane potential on sodium conductance in the giant axon of *Loligo*. *J. Physiol., Lond.*, **116** (1952), 497–506

31. A.L. Hodgkin and A.F. Huxley. A quantitative description of membrane current and its application to the conduction and excitation in nerve. *J. Physiol., Lond.*, **117** (1952), 500–44

32. A.L. Hodgkin. The Croonian Lecture: Ionic movements and electrical activity in giant nerve fibres. *Proc. R. Soc. B,* **148** (1958), 1–37

33. N.B. Standen. Properties of a calcium channel in snail neurones. *Nature, Lond.*, **250** (1974), 340–2

34. S. Weidmann. Heart: electrophysiology. *A. Rev. Physiol.*, **36** (1974), 155–69

35. H.D. Lux and R. Eckert. Inferred slow inward current in snail neurones. *Nature, Lond.*, **250** (1974), 574–6

36. B. Hille. Ionic channels in nerve membranes. *Prog. Biophys. molec. Biol.*, **21** (1970), 1– 32

37. B. Hille. The permeability of the sodium channel to metal cations in myelinated nerve. *J. gen. Physiol.*, **59** (1972), 637–58

38. B. Hille. Potassium channels in myelinated nerve. Selective permeability to small cations. *J. gen. Physiol.*, **61** (1973), 669–86

39. A.M. Woodhull. Ionic blockage of sodium channels in nerve. *J. gen. Physiol.*, **61** (1973), 687–708

40. C.M. Armstrong, F. Benzanilla and E. Rojas. Destruction of sodium conductance inactivation in squid axons perfused with pronase. *J. gen. Physiol.*, **62** (1973), 375–91

41. P.G. LeFevre. Sugar transport in the red blood cell: structure–activity relationships in substrates and antagonists. *Pharmacol. Rev.*, **13** (1961), 39–70

42. P.G. LeFevre and G.F. McGinniss. Tracer exchange vs net uptake of glucose through human red cell surface. New evidence for carrier-mediated diffusion. *J. gen. Physiol.*, **44** (1960), 87–103

43. C.R. Park, R.L. Post, C.F. Kalman, J.H. Wright, H.L. Johnson and H.E. Morgan. The transport of glucose and other sugars across cell membranes and the effect of insulin. *Ciba Fdn. Colloq. Endocr.*, Vol. **9** (1956), 240–60

44. M. Levine, D.L. Oxender and W.D. Stein. The substrate-facilitated transport of the glucose carrier across the human erythrocyte membrane. *Biochim. Biophys. Acta*, **109** (1965), 151–63

45. P. Mitchell. Chemiosmotic coupling in oxidative and photosynthetic phosphorylation. *Biol. Rev.*, **41** (1966), 445–502

46. O.H. LeBlanc. Effect of uncouplers of oxidative phosphorylation on lipid bilayer membranes: carbonylcyanide *m*-chlorophenylhydrazone. *J. Membr. Biol.* **4** (1971), 227–51

47. A. Finkelstein. Weak-acid uncouplers of oxidative phosphorylation. Mechanism of action on thin lipid membranes. *Biochim. Biophys. Acta*, **205** (1970), 1–6

48. M. Foster and S. McLaughlin. Complexes between uncouplers of oxidative phosphorylation. *J. Membr. Biol.*, **17** (1974), 155–80

49. E.P. Bakker, E.J. Van Den Heuvel, A.H.C.A. Wiechmann and K. Van Dam. A comparison between the effectiveness of uncouplers of oxidative phosphorylation in mitochondria and in different artificial membrane systems. *Biochim. Biophys. Acta*, **292** (1973), 78–87

50. P.J.F. Henderson, J.D. McGivan and J.B. Chappell. The action of certain antibiotics on mitochondrial, erythrocyte and artificial phospholipid membranes. *Biochem. J.,* **111** (1969), 521-35

51. R.C. Prince, A.R. Crofts and L.K. Steinrauf. A comparison of beauvericin, enniatin and valinomycin as calcium transporting agents in liposomes and chromatophores. *Biochem. Biophys. Res. Comm.,* **59** (1974), 697-703

52. S. Roseman. Carbohydrate transport in bacterial cells. In: *Metabolic Pathways,* vol. 6 (L.E. Hokin ed.), Academic Press, New York (1972), pp 41-89

53. W. Kundig and S. Roseman. Sugar transport, I isolation of a phosphotransferase system from *Escherichia coli. J. biol. Chem.,* **246** (1971), 1393-406

54. W. Kundig and S. Roseman. Sugar transport, II characterization of constitutive membrane-bound enzymes II of the *Escherichia coli* phosphotransferase system. *J. biol. Chem.,* **246** (1971), 1407-18

55. R.D. Simoni, T. Nakazawa, J.B. Hays and S. Roseman. Sugar transport, IV isolation and characterization of the lactose phosphotransferase system in *Staphylococcus aureus. J. biol. Chem.,* **248** (1973), 932-40

56. J.B. Hays, R.D. Simoni and S. Roseman. Sugar transport, V A trimeric lactose-specific phosphocarrier protein of the *Staphylococcus aureus* phosphotransferase system. *J. biol. Chem.,* **248** (1973), 941-56

57. R.D. Simoni, J.B. Hays, T. Nakazawa and S. Roseman. Sugar transport, VI phosphoryl transfer in the lactose phosphotransferase system of *Staphylococcus aureus. J. biol. Chem.,* **248** (1973), 957-65

58. R.D. Simoni and S. Roseman. Sugar transport, VII lactose transport in *Staphylococcus aureus. J. biol. Chem.,* **248** (1973), 966-76

59. S.G. Schultz and P.F. Curran. Coupled transport of sodium and organic solutes. *Physiol. Rev.* **50** (1970), 637-718

60. P.F. Curran, S.G. Schultz, R.A. Chez and R.E. Fuisz. Kinetic relations of the Na-amino acid interaction at the mucosal border of intestine. *J. gen. Physiol.,* **50** (1967), 1261-86

61. J.F. Schaeffer, R.L. Preston and P.F. Curran. Inhibition of amino acid transport in rabbit intestine by *p*-chloromercuriphenyl sulfonic acid. *J. gen. Physiol.,* **62** (1973), 131-46

62. G.A. Kimmich. Coupling between Na^+ and sugar transport in small intestine. *Biochim. Biophys. Acta,* **300** (1973), 31-78

63. J.A. Schafer and E. Heinz. The effect of reversal of Na^+ and K^+ electrochemical potential gradients on the active transport of amino acids in Ehrlich ascites tumor cells. *Biochim. Biophys. Acta,* **249** (1971), 15-33

64. H.N. Christensen, C. De Cespedes, M.E. Handlogten and G. Ronquist. Energization of amino acid transport, studied for the Ehrlich ascites tumor cell. *Biochim. Biophys. Acta,* **300** (1973), 487-522

65. R. Chesney, B. Sacktor and A. Kleinzeller. The binding of phloridzin to the isolated luminal membrane of the renal proximal tubule. *Biochim. Biophys.*

Acta, **332** (1974), 263- 77

66. M. Silverman. The chemical and steric determinants governing sugar interactions with renal tubular membranes. *Biochim. Biophys. Acta*, **332** (1974), 248–62

67. F. Fuchs. Striated muscle. *A. Rev. Physiol.*, **36** (1974), 461-502

68. D.A. Thorley-Lawson and N.M. Green. Studies on the location and orientation of proteins in the sarcoplasmic reticulum. *Eur. J. Biochem.*, **40** (1973), 403–13

69. P.S. Stewart and D.H. MacLennan. Surface particles of sarcoplasmic reticulum membranes. Structural features of adenosine triphosphatase. *J. biol. Chem.*, **249** (1974), 985–93

70. C.F. Louis, R. Buonaffina and B. Binks. Effect of trypsin on the proteins of skeletal muscle sarcoplasmic reticulum. *Archs. Biochem. Biophys.* **161** (1974), 83–92

71. A. Martonosi. Biochemical and clinical aspects of sarcoplasmic reticulum function. *Curr. Top. Membr. Transp.*, **3** (1972), 83–197

72. G. Inesi, M. Millman and S. Eleter. Temperature-induced transitions of function and structure in sarcoplasmic reticulum membranes, *J. molec. Biol.*, **81** (1974), 483–504

73. H. Bader, R.L. Post and G.H. Bond. Comparison of sources of a phosphorylated intermediate in transport ATP'ase. *Biochim. Biophys. Acta*, **150** (1968), 41–6

74. J.C. Ellory and V.L. Lew. A K^+-dependent phosphatase in the membranes of low-K^+-type erythrocytes. *Biochim. Biophys. Acta*, **332** (1974), 215–20

75. J.J.H.H.M. De Pont, A. Van Prooijen-Vaneeden and S.L. Bonting. Studies on (Na^+-K^+)-activated ATPase XXXIV. Phosphatidylserine not essential for (Na^+-K^+)-ATPase activity. *Biochim. Biophys. Acta*, **323** (1973), 487–94

76. R. Whittam and M.E. Ager. Vectorial aspects of adenosine-triphosphatase activity in erythrocyte membranes. *Biochem. J.*, **93** (1964), 337–48

77. P.F. Baker. Phosphorus metabolism of intact crab nerve and its relation to the active transport of ions. *J. Physiol., Lond.*, **180** (1965), 383–423

78. V.T. Marchesi and G.E. Palade. The localization of Mg–Na–K-activated adenosine triphosphatase on red cell ghost membranes. *J. Cell Biol.*, **35** (1967), 385–404

5 Regulation of Cellular Processes by Membranes

The outer surface of the plasma membrane serves as a repository for the various molecular components required by the cell to receive and respond appropriately to external stimuli. Other factors are also present on the membrane surface which establish the indigenous identity of the cell and assist in immunological defence reactions. Intracellular membranes are not exposed to the external environment and, in general, they do not possess cell-surface-specific components. These membranes, which are particularly well developed in most eukaryotic cells, create a system of compartments within the cytoplasm which bring about a segregation of the various subcellular activities. This is accomplished, in part, by directing the transfer of materials from one site to another within the cell as well as regulating the transposition of substances from one compartment to another. Furthermore, enzymic reactions may be confined to specified membranes located at strategic sites within the cell and the membrane itself may also provide a suitable microenvironment for these reactions to take place.

5.1 The Plasma Membrane: Surface Mediated Cellular Responses

Certain components exposed at the surface of the plasma membrane fulfil a functional role concerned with the interaction of the cell with surrounding cells and with body fluids. These interactions, at least in higher organisms, are essential in an organisational sense to direct the association of cells into tissues and organs and subsequently to mediate their affiliated social behaviour. The presence of specific cell-recognition sites on the plasma membrane has been demonstrated by Merrell and Glaser[1] in studies of embryonic neural retina and cerebellar tissue. Dissociation of these tissues can be accomplished by mild proteolytic and other treatments to yield suspensions of intact cells, which eventually reaggregate to reproduce many of the features of the original tissue. When cells from different origins are mixed, the two populations reassociate into separate homotypic aggregates, indicating some form of recognition between cells. The specificity of these sites and their location on the surface of the plasma membrane was inferred from experiments in which plasma membrane fragments from these and other cells were mixed with neural retina or cerebellar cell suspensions. It was found that reaggregation is prevented only when homotypic membrane fragments are added but not when plasma membrane from any other source is supplied. Apart from adhesion and communication between specific cells, another important feature associated with contact between cells is that growth comes under much stricter control. This phenomenon is referred to as density-dependent inhibition, and one of the most obvious signs of malignancy in cells is the loss of this control. The plasma membrane is also involved in locomotion and chemotaxis, enabling cells to move from an unfavourable to a more favourable environment.

Apart from innervation, the main outside control exercised over specific cellular functions is that mediated by hormones. These substances interact with specific membrane receptors located on the cell surface through which they are able to initiate secondary events within the cell. Other cellular responses can also be induced by the interaction of various antigens with membrane receptor sites and these are important in the development of immune responses. Numerous attempts have been made to characterise particular surface receptor sites but these efforts are often frustrated by problems concerned with isolating and identifying surface components from membranes in an active form. The general conclusion from such studies, however, is that the molecular structure of most of these sites is highly complex and depends to a certain extent on interaction with other membrane constituents. Specific interactions between cellular effectors and their corresponding receptor sites is believed to arise from regions of complementary conformation on the respective molecules; a lock and key analogy is often invoked to describe this concept. Receptor-site specificity in this context is regarded as an exclusive binding of a certain molecule to a membrane receptor site and, in intact systems, a particular response by the cell can often be observed.

5.1.1 Hormone Receptor Sites and Adenylcyclase

The action of many hormones is mediated through activation of the membrane-bound enzyme, adenylcyclase. This enzyme catalyses the synthesis of cyclic 3′, 5′-AMP from ATP and the product serves as a second messenger within the cell. Enzyme activity has been demonstrated in a wide variety of animal cells and tissues as well as yeasts and fungi and, although it has not yet been detected in higher plants, this possibility cannot be excluded at the present time. The enzyme was originally thought to be confined exclusively to the plasma membrane, but it has been found recently in nuclear membranes, although its function here is a matter of conjecture. Cyclic AMP effects certain key reactions in different cell types and a delicate balance of its concentration within the cytoplasm appears to be an essential ingredient in cellular control mechanisms. This balance is achieved by the action of a phosphodiesterase enzyme, which rapidly converts the cyclic phosphodiester to AMP, and this enzyme invariably operates in conjunction with adenylcyclase. The combined action of these two enzymes serves to maintain the intracellular cyclic AMP concentration within limits appropriate to the prevailing functional requirements of the cell.

The detailed molecular configuration of adenylcyclase is unknown, but Rodbell[2] has presented evidence which indicates that the enzyme is likely to be a composite of at least two separate parts; a receptor site and a catalytic site. The catalytic site is probably the same in all cells since all involve the conversion of ATP to cyclic AMP and PP, but the enzyme can only be activated, under normal conditions, by interaction of the appropriate hormone with the particular receptor site. Specific membrane receptors have been described for various polypeptide hormones including thyroid-stimulating hormone, adrenocorticotrophic hormone, leuteinising hormone, melanocyte-stimulating hormone, parathyroid hormone, glucagon, secretin and vasopressin. All these hormones have been shown to stimulate membrane-bound adenylcyclase of respective target cells through a specific binding to receptor sites on the plasma membrane. Other compounds such as catecholamines and prostaglandins are also able to stimulate adenylcyclase but it is not clear whether the receptor sites for these compounds are the same type as those for the polypeptide hormones. Some cells are known to possess several hormone-specific receptor sites for adenylcyclase, each of which can act independently of or, in some instances, simultaneously with others. Plasma membranes of epididymal fat-pad cells of the rat, for example, possess discrete binding sites for six polypeptide hormones, each of which can stimulate adenylcyclase activity. It is apparent that under these conditions there is no direct discrimination in subcellular response between the different primary effectors.

5.1.2 Orientation of Adenylcyclase in the Plasma Membrane

Adenylcyclase acts as the vehicle by which hormones can direct events within the cell without having to enter it. Obviously the enzyme complex must be orientated within the membrane so that the receptor site is accessible to the hormone on the outer surface of the plasma membrane while the catalytic site must deliver the reaction product to the cytoplasm. A diagrammatic representation of the probable relationship between the adenylcyclase complex and the plasma membrane is shown in figure 5.1. The external aspect of the receptor site is indicated by the fact that hormone binding can be abolished by treatment of intact cells with trypsin; enzymic hydrolysis of the membrane constituents in this case is restricted solely to proteins exposed on the outer surface of the membrane. Hormones can also be displaced from the receptor site by adsorption to antibodies prepared against specific hormones and, as with trypsin, the size of the antibody molecule prevents its penetration through the plasma membrane. A novel, and more direct, approach to this problem has been developed recently whereby hormones are covalently bound to high relative molecular mass sugar polymers. Complexes have been prepared with adrenocorticotrophic hormone and glucagon, both of which can activate adenylcyclase when added to a suspension of intact cells. The evidence for assigning the catalytic site of adenylcyclase to the cytoplasmic surface of the membrane comes from two sorts of experiments. Firstly, the nucleotide substrate for the reaction (ATP) must be supplied to the inner surface of the membrane before hormone activation can occur; substrate added outside the cell is not converted into cyclic AMP unless the membrane is rendered permeable to ATP. The second line of evidence concerns the susceptibility of adenylcyclase activity to the action of proteolytic enzymes such as trypsin and pepsin. The basal or unstimulated activity of adenylcyclase is not diminished when the action of these enzymes is confined to the outer surface of the plasma membrane but, when digestion of the cytoplasmic surface is permitted, adenylcyclase activity is completely abolished.

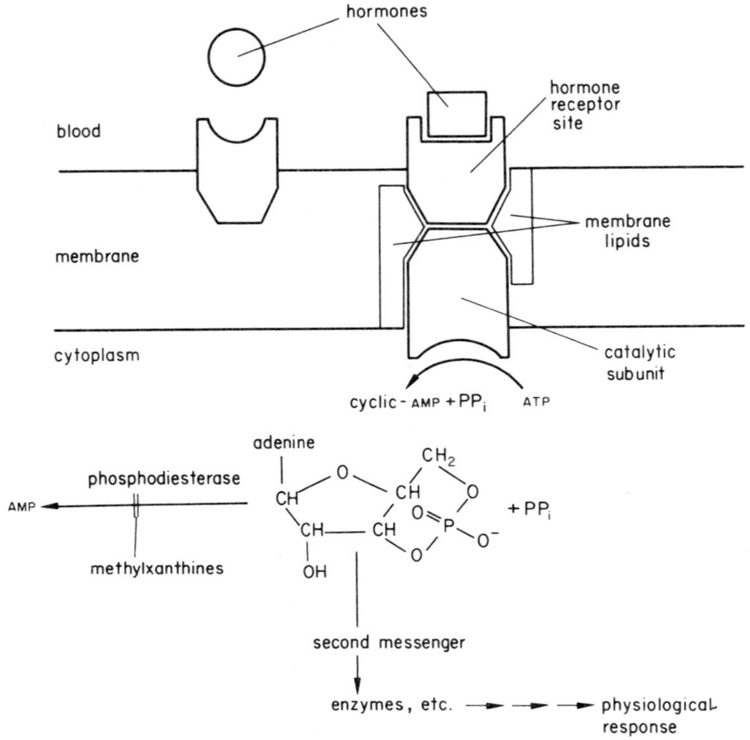

Figure 5.1 Components of the adenylcyclase enzyme complex. Hormones (primary messengers) bind to specific-membrane receptor sites (discriminators) located on the outer surface of the plasma membrane. Binding studies have indicated that there are more receptor sites than catalytic sites in some membranes and hormone-binding alone will not activate the enzyme unless the binding complex is associated with the catalytic site. The catalytic subunit is located in the membrane with the active centre accessible to the cytoplasm. Specific-membrane lipids associated with either or both receptor and catalytic components may assist in the orientation of the activated complex or they may even transduce the hormone-binding signal to the catalytic subunit if no direct protein–protein interactions are involved. Fluoride and prostaglandins are believed to operate at this level in bringing about an activation of the enzyme. The reaction at the catalytic site is the formation of the second messenger compound cyclic AMP from ATP. Cyclic AMP mediates a variety of reactions in the cell, which contribute to the overall physiological response to the initial hormone-binding. Cyclic AMP is converted to 5′AMP by a specific phosphodiesterase that can be inhibited by methyl xanthines.

Activation of adenylcyclase is found to be more complex than a simple binding of the hormone to a receptor site since in several instances additional factors are required to maximise hormone stimulation. The nucleotide GTP, for example, has been shown to augment adrenaline and glucagon stimulation of liver adenylcyclase and the activation of platelet adenylcyclase by prostaglandin E_1, whereas calcium is required for enzyme activation by melanocyte-stimulating hormone and adrenocorticotrophic hormone in adrenal membranes. These cofactors are believed to exert an allosteric control over hormone-binding to the receptor site although it is possible they may operate at another level in the stimulation sequence. A transducer element is considered by some workers to constitute a separate component of the enzyme complex and it is thought that this component is interposed between the receptor and catalytic sites and is responsible for transmitting the binding signal vectorally across the membrane to the catalytic site[3]. Although this third component has not yet been identified, the action of a number of agents that stimulate enzyme activity are consistent with an effect at a site other than the hormone receptor site. Prostaglandins and fluoride, for example, stimulate most hormone-specific adenylcyclase enzymes and this stimulation is not additional to that which can be induced by the particular hormone concerned. Differences in the mode of stimulation of adenylcyclase by hormones and other agents has been examined by Kreiner *et al.*[4] in studies of the temperature-sensitive changes in enzyme activity associated with liver plasma membranes (figure 5.2). They found an abrupt increase in enzyme stimulation by adrenaline and glucagon at 32°C, which amounted to a doubling of the energy of activation. No inflection point was observed either in basal activity or in fluoride or prostaglandin E_1 activation, indicating that the site of stimulation by these agents was independent of the temperature-sensitive hormone site. The differences in hormone activation above and below 32°C did not represent differences in hormone or substrate binding so they concluded that the effect must involve an interaction between components of the enzyme complex. Because of the abrupt nature of the change in activation energy with temperature, lipid phase structure was considered to be of possible importance in the hormone-stimulation sequence, but an effect restricted to protein components alone could not be excluded. Some evidence that membrane lipids are involved

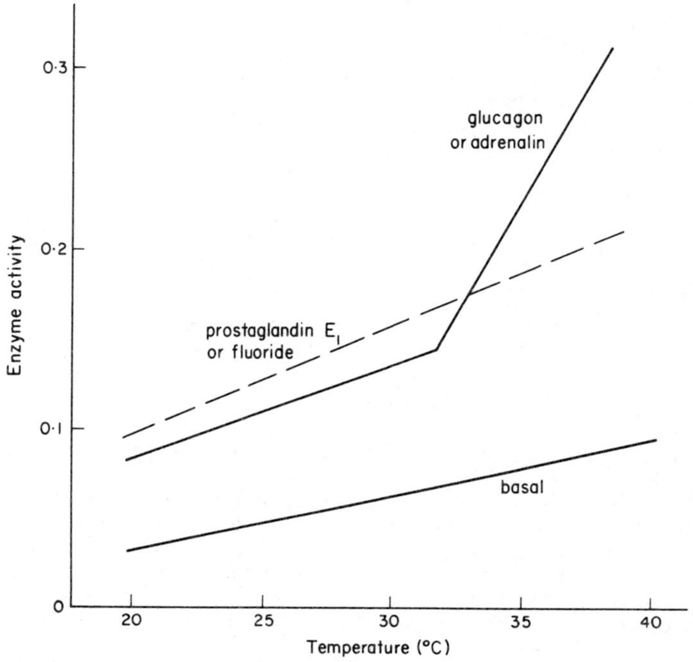

Figure 5.2 The stimulation of adenylcyclase of liver-plasma membranes by hormones and other agents as a function of temperature. Enzyme activity is expressed in nanomoles of cyclic AMP per minute per milligram membrane protein (adapted from reference 4).

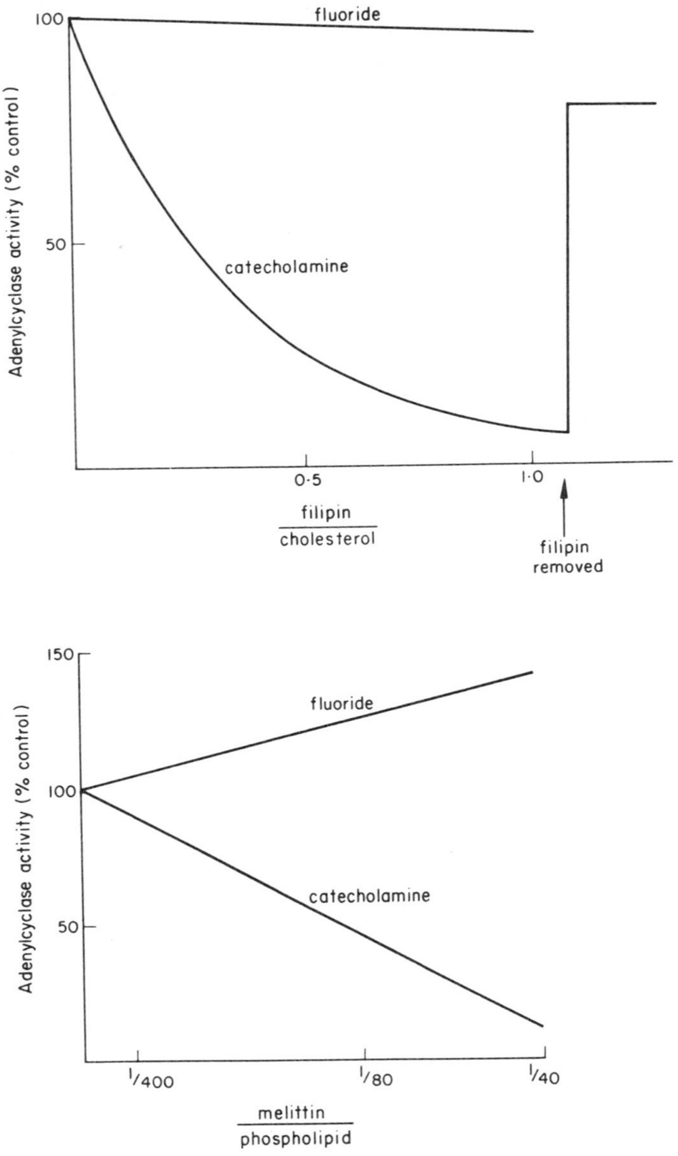

Figure 5.3 Uncoupling of hormone activation of adenylcyclase by agents that interact with membrane lipids.

(a) Addition of the polyene antibiotic, filipin, to avian erythrocyte membranes, in amounts that do not disrupt the membrane, removes catecholamine activation of adenylcyclase. Filipin treatment does not affect hormone-binding and adenylcyclase is again activated by catecholamine when filipin is removed from the membrane. Fluoride activation of adenylcyclase is not affected by the presence of filipin. The antibiotic interacts specifically with membrane cholesterol, but only some of this lipid is involved in the coupling mechanism.

(b) Melittin from bee venom, a peptide of twenty-six amino acid residues (of which twenty are hydrophobic or neutral in character), interacts with the phospholipid component of avian erythrocyte membranes, also causing a loss of catecholamine activation of adenylcyclase. Fluoride activation is slightly enhanced by the presence of melittin (from reference 5).

in modulating catalytic activity by hormone-binding has been obtained by Puchwein *et al.*[5] who showed that agents that interact specifically with either cholesterol or phospholipids could almost completely abolish catecholamine activation of adenylcyclase (see figure 5.3). The loss in hormone stimulation, at least in the case of a cholesterol-filipin complex, did not appear to result from a decrease in hormone-binding to the receptor but was more consistent with an uncoupling of the hormone–receptor complex from the catalytic site. In experiments where filipin was subsequently extracted from the membrane, catecholamine activation could be restored to near normal levels, indicating that the effect was completely reversible and that no selective destruction of the receptor site occurred. Stimulation of the enzyme by fluoride when lipids were complexed with filipin or melittin was either unaffected or showed a modest increase in stimulation, confirming the notion that the site of fluoride activation is not connected to the hormone-receptor site.

5.1.3 Interaction of Adenylcyclase with Membrane Lipids

Adenylcyclase is a lipoprotein, which relies on interaction with other membrane components to maintain it in an active configuration and responsive to hormones and other agents. Membrane phospholipids appear to be important in this respect, since treatment of plasma membranes with phospholipase A or C reduces basal activity of the enzyme and abolishes hormone activation[6]. Furthermore, treatment of membrane preparations from brain with ether, a procedure that preferentially extracts the least polar lipids, has been shown to have a minimal effect on hormone activation of adenycyclase, whereas selective extraction of the more polar phospholipids from liver membranes with an ether:butanol solvent almost completely abolishes hormone activation of the enzyme and substantially reduces the basal adenylcyclase activity as well.

Levey[7] has succeeded in achieving an almost complete extraction of adenylcyclase from heart-muscle tissue with nonionic detergent and the enzyme was found to be a protein complex with an estimated relative molecular mass between 100 000 and 200 000. The solubilised enzyme could be stimulated by fluoride but was no longer responsive to catecholamines, glucagon or thyroxine, although these hormones were still able to bind to the complex, suggesting that the interaction between the receptor and catalytic components was partially disrupted during the extraction procedure. Levey was able to demonstrate a restoration of basal enzyme activity, more than half of which was lost on removal of detergent by ion-exchange chromatography, as well as responsiveness to glucagon and adrenalin on addition of certain phospholipids. The specificity of phospholipids concerned in basal and hormone-responsive activities of adenylcyclase has been examined by Rethy *et al.*[8] in phospholipase-treated and solvent-extracted liver-plasma membranes. They showed that ether:butanol extraction reduces basal adenylcyclase activity and responsiveness to hormones and fluoride but that both activities could be restored by adding specific phospholipids to the extracted membranes. Thus, complete recovery of basal activity was obtained in the presence of phosphatidylinositol but hormone or fluoride activation was not affected. Almost complete restoration of adrenalin activation and the recovery of some glucagon and fluoride sensitivity could be brought about by the additon of phosphatidylserine. These and other observations suggest that interaction of specific membrane phospholipids with different components of the adenylcyclase enzyme is required in the formation of an active hormone–receptor complex and also to modulate the basal activity of the catalytic site.

5.1.4 Other Hormone Receptor Sites on the Plasma Membrane

A notable exception to the list of polypeptide hormones that stimulate adenylcyclase is insulin. This hormone does not stimulate cyclic AMP production: on the contrary, it appears to act in certain circumstances by reducing the intracellular concentration of cyclic AMP. Insulin, for example, is reported to antagonise the lipolytic action of moderately stimulating concentrations of a number of polypeptide hormones and catecholamine when applied to epidymal fat-pad cells and it appears to do this by reducing the amount of cyclic AMP in the cell. This action of insulin is thought to involve an activation of phosphodiesterase, which catalyses the hydrolysis of cyclic AMP to AMP, since an increase in phosphodiesterase activity in some tissues can be correlated with the antagonistic effect of insulin on adenylcyclase stimulation by the other hormones. The question of insulin antagonism through a direct effect on the phosphodiesterase activity, however, is by no means resolved, especially since this would require the entry of insulin into the cell. The possibility that insulin may act by stimulating the production of a different intracellular messenger compound, similar to cyclic AMP, cannot therefore be excluded. In other cells such as liver, insulin in physiological concentrations has been shown to prevent a rise in intracellular cyclic AMP concentration by acting directly to inhibit hormonal activation of adenylcyclase.

Apart from its effect on intracellular cyclic AMP concentration, insulin initiates a multiplicity of changes in cellular metabolism in susceptible tissues[9]. Many of these changes are thought to reflect an alteration in intracellular pool size of amino acids, sugars and other compounds whose transport into the cell is augmented by the hormone. Nevertheless, some effects of insulin cannot be adequately explained simply in terms of changes in membrane permeability. Stimulation of protein biosynthesis, for example, appears to be brought about by an increase in the amount of ribosomal RNA rather than by an enlargement of the amino acid pool size. Furthermore, there is some evidence to show that exogenous amino acids taken up by cells treated with insulin are preferentially incorporated into protein, and not those of the intracellular pool, suggesting that the hormone may have another effect on protein biosynthesis elsewhere in the cell (see section 5.1.5).

Steroid hormones are exceptional in that they are not known to have receptor sites located in the plasma membrane. Receptor sites for these hormones appear to be mainly soluble proteins located within the cell although a site of action on other cell membranes has been demonstrated (see section 5.2.2). The hydrophobic character of steroid hormones is likely to facilitate their penetration of the plasma membrane, in contrast to polypeptide hormones, which have an extracellular *modus operandi*.

5.1.5 The Insulin Receptor Site

Since changes in membrane permeability are among the earliest insulin effects observed, it is likely that the initial site of insulin action is at the surface of the plasma membrane. This has been confirmed by studies using hormone covalently bound to high molecular mass polymers which can induce typical insulin responses in susceptible tissues even though the complex is large enough to preclude entry of insulin into the cell[10]. The presence of a specific insulin receptor site has been investigated by a number of workers[11] using a radioactive iodine derivative of insulin which retains full biological activity. In general it has been found that insulin can bind non-specifically to membranes and even to cells that do not normally respond to insulin, but a certain proportion does bind with saturation kinetics to plasma membranes of responsive cells. Insulin bound specifically to the membrane can be displaced with native hormone and binding does not appear to involve any chemical change in the hormone or receptor site. Insulin-binding, which itself is not markedly temperature sensitive, induces certain temperature-sensitive changes in the membrane that are believed to cause the observed physiological responses such as enhanced sugar or amino acid transport.

There has been some progress lately in characterising the insulin receptor site. The receptor appears to be a protein since specific insulin-binding is abolished by digestion of surface-membrane proteins with trypsin, and the binding protein can be extracted from membranes with detergent. When partially purified, the detergent-solubilised receptor complex has a relative molecular mass of more than 300 000 and is able to bind insulin to the same extent as the intact membrane site. The receptor protein has a carbohydrate component but there is no indication at this stage to indicate that these residues are concerned with binding specificity. The sialic acid component at least can be eliminated because treatment of membranes with neuraminidase (which specifically removes sialic acid residues) doe not appear to affect insulin-binding. Whether or not membrane phospholipids are involved in insulin-binding or subsequent changes in the membrane associated with binding is not clear. Phospholipase treatment of certain membranes actually increases specific insulin-binding four to six times, suggesting that there is a reserve of cryptic insulin receptor sites, which are masked by phospholipids, but studies involving proteolytic destruction of insulin receptors of the fat-cell membrane have indicated a rapid reappearance of specific binding sites during incubation after the initial enzymic treatment. These new receptor sites are not likely to arise from the expression of previously masked insulin-binding sites in the membrane because their appearance depends on active protein synthesis, so presumably they are newly synthesised receptor proteins which become inserted into the plasma membrane. Binding of insulin to the receptor site alone, of course, does not imply that a signal will be transduced in the membrane and there is some evidence that membrane phospholipids are required to couple insulin-binding with processes concerned with transport of substances across the membrane.

5.1.6 Histocompatibility Antigens

Cells of all mammalian species possess antigens located on the surface of the plasma membrane, which appear to function primarily in recognition. These antigens or determinants, as they are known, are also thought to play an important role in other, hitherto unidentified, membrane phenomena such as density-dependent inhibition of cell growth and induction of mitotic division (mitogenesis) in cells in a resting state. The particular determinants involved in cell recognition are known as the histocompatibility antigens[12]. They can be identified readily by several immunological techniques, including allografting. This procedure involves grafting tissue from one individual to another within a particular species enabling differences between strains to be recognised on the basis of acceptance or rejection of the graft by the host. The determinants of each species so far examined are derived from closely linked genes situated in a single region of the chromosome in a position known as the histocompatibility locus. These genes are responsible for formulating the production of antigen or, in some species, a series of similar or homologous antigens. The histocompatibility antigens of several species have been examined in some detail including the mouse, guinea pig, rat, chicken and man, and it is likely that most other species possess similar antigens. Serological tests performed on cells from the above species have shown that the histocompatibility antigens can be subdivided into a number of different and unique specificities. The human histocompatibility antigens (HL–A), for example exhibit more than twenty immunologically distinct specificities arising from different gene combinations, although many of these are difficult to detect by standard immunological tests and are said to be weakly expressed[13]. The chemical composition of different (HL–A) antigens isolated from human plasma membranes are found to be similar and some parts of the molecule are apparently identical. The functional groups of the antigen are thought to be confined to a certain region of the molecule and then only small changes may be required to produce differences in antigenic specificity. Histocompatibility antigens can also appear as a result of infection of host cells by a virus. Human leukaemic cells, for example, express other distinctive histocompatibility antigens (referred to as tumour-specific transplantation antigens) in addition to the normal HL–A antigens. The new determinants are believed to arise by viral induction from the host chromatin or alternatively are coded directly by the viral genome.

5.1.7 Blood-group Determinants

Another specialised group of plasma-membrane antigenic determinants are the blood-group substances, and in man at least fourteen different systems have been described, which are known to express more than sixty different blood-group specificities[14]. These mosaic structures are believed to be products of one or more genes and their respective allelomorphs and include the AB and MN determinants and related Lewis substances. Many of these antigens are not restricted to the erythrocyte membrane but are also found in a soluble form in various body fluids and secretions. The soluble antigens are usually indistinguishable, both chemically and immunologically, from the respective antigens bound to the plasma membrane, however. It is unlikely that the two forms are readily interconvertible because the bound antigens are firmly attached to the membrane by hydrophobic interactions. These antigens require detergent action to extract them from the plasma membrane and, when solubilised, they can be recovered in an immunocompetent form.

All the main blood-group substances appear to be glycoproteins with antigenic specificity determined predominantly by the carbohydrate residues. The Rh antigen of human erythrocytes, however, is an exception to this rule, since it is a protein with no carbohydrate component and it requires oxidised sulphydryl groups for antigenic expression. The Rh antigen is difficult to extract from the membrane in an active form and other membrane components seem to be involved in the antigenic specificity, possibly explaining why the antigen is so labile. Specific interaction between the Rh antigen and certain phospholipids has been reported by Green[15], who showed that both polar and hydrophobic regions of the phospholipid molecules participate in the binding. Phosphatidylcholine and phosphatidylethanolamine were the only phospholipids examined that were capable of restoring Rh antigenic activity to the purified protein and at least one unsaturated fatty acyl residue was necessary in order to reconstitute the active antigen; phospholipids with completely saturated hydrocarbon chains did not interact specifically with the antigen.

5.1.8 Surface Receptors and Immune Responses

Antibody receptor sites of unique specificity residing in the plasma membrane of lymphocytes are involved in the development of immune responses in the body. The existence of two populations of antigen-reactive lymphocytes, both originating in the bone marrow of adults, has been described[16,17] (see figure 5.4). One class of cells migrate to the thymus and acquire the ability to respond to antigen (thymus or T-cells) and another class constitute the progenitors of mature antibody-secreting cells (bursa or B-cells). T-cells are responsible for what are known as cell-mediated immune responses, including delayed sensitivity, homograft and graft–host reactions. In the mouse, these cells are characterised by the presence of a specific antigen located on the plasma membrane surface (θ antigen) although surface immunoglobulin has also been detected. This immunoglobulin however, does not appear to be produced by T-cells, nor does it constitute a specific antigen receptor site. Thus, it is probably adsorbed non-specifically to the plasma membrane. Cell-mediated immune responses are initiated when a subpopulation of T-cells recognise and bind circulating foreign antigen, which is the signal for these cells to become activated and proliferate by mitotic division. The primed cells can then participate in specific cytotoxic reactions (see section 5.1.9).

In addition to the T-cell-mediated immunity there is a second loop of the immune response called humoral immunity in which the B-cells are the primary effectors. The B-cells, unlike T-cells, are capable of synthesising and secreting antibody, and the antigen receptor sites on the surface of these cells are membrane-bound immunoglobulin molecules. B-cells can be stimulated directly by

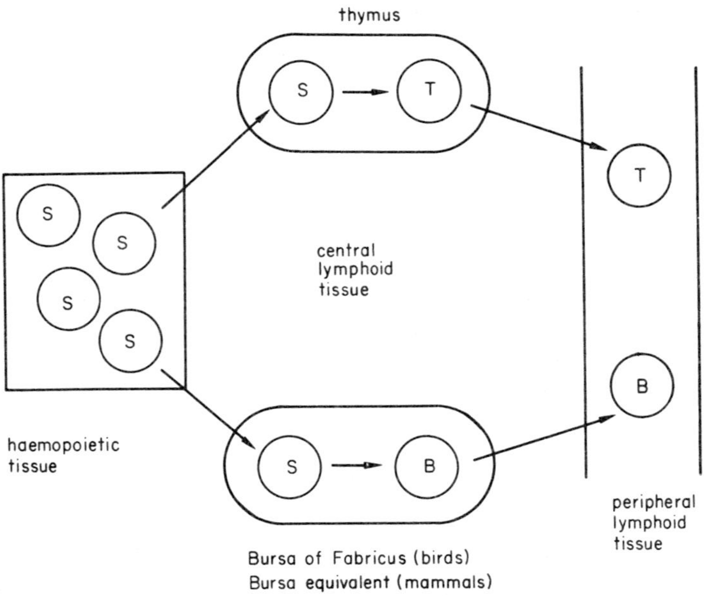

Figure 5.4 Diagram of the pathwyas of differentiation of lymphocytes concerned with immune reactions. Undifferentiated stem cells (S) from the yolk sac and liver of the foetus or bone marrow of adults migrate to the thymus, where they differentiate into T-cell lymphocytes (T). Stem cells migrating to the Bursa of Fabricus in birds or its equivalent in mammals differentiate into B-cell lymphocytes (B). T-cells and B-cells both migrate to the peripheral lymph nodes, gut-associated lymph tissues and the spleen, where they participate in immune reactions (adapted from reference 16).

specific antigen-binding to these receptor sites and such cells respond by producing a certain class of immunoglobulin molecules directed against the bound antigens. Although direct stimulation of B-cells by antigen does occur in the development of the humoral immunity it is not accepted as the usual method of B-cell activation. The principal mechanism appears to be mediated by T-cells that have been activated by antigen. The mechanism of interaction between T-cells and B-cells in the aetiology of humoral immunity is not known but two plausible theories have been considered. The first involves a migration of T-cells, activated by specific antigens, to regions of the lymphatic system that are heavily populated with B-cells. Different antigenic determinants on the B-cell membrane are thought to recognise antigen, possibly presented as a multi-determinant antigen array on the T-cell surface, and produce antibody against this antigen. The mediation of B-cell activation in the alternative hypothesis is believed to occur by an interaction with soluble, nonimmunogenic messenger compounds, which are secreted by activated T-cells. Attempts have been made to isolate and characterise such factors, but as yet these have not been successful[18]. Whatever the mechanism of communication between lymphocytes, the obvious advantage of an interdependence between the T-cells and B-cell activation is that differentiation of B-cells in response to trivial antigens is thus prevented.

5.1.9 Membrane-mediated Cytotoxic Mechanisms

The plasma membrane is also the primary site concerned in cytoxic mechanisms associated with immune responses. Cell-mediated immunity is characterised by the rapid destruction of large numbers of specific cells by activated T-cell lymphocytes, which establish contact with target cells through an interaction between membrane-bound cell-specific antigens and receptors. Cytotoxic substances produced by the activated T-cells are thought to be transmitted directly to target cells following agglutination because such factors are never detected in the medium surrounding these cells. Death of the target cell may follow immediately contact is established, but occasionally there is a delay and the T-cell lymphocyte may even become detached before the target cell succumbs. Cell-mediated cytotoxic mechanisms also involve co-operative interactions between T-cells and macrophages that phagocytose target cells[19]. The role of macrophages, however, is not restricted solely to phagocytosis, since it appears that they are capable of killing cells by the production of cytotoxic factors. Temple *et al.*[20], for example, have shown that destruction of target cells is enhanced when phagocytic activity of macrophages is suppressed, and they suggested that this was accomplished by the transfer of lysosomal enzymes from macrophages to target cells when the plasma membranes of the two were in close contact.

Cytotoxicity associated with humoral immunity is brought about by the creation of holes in the plasma membrane of the target cells through which the cytoplasmic contents of the cell can escape. This process, known as complement-mediated lysis, involves interaction between antigenic sites located on the plasma membrane of the target cell and complementary antibodies secreted by differentiated B-cells. The cytotoxic process that follows this initial reaction has been described in detail by Müller-Eberhard[21] (figure 5.5). Briefly, there are nine separate protein complement factors, concerned in the lytic mechanism designated C_1 to C_9, produced by the liver, macrophages and other cells and all are present in serum. The first complement factor, (C_1), consisting of a calcium-dependent complex of three subunits, has six immunoglobulin-binding sites. At least two membrane-bound immunoglobulin molecules in close proximity are required for C_1 binding. This complex of antigen–antibody and C_1 (figure 5.5a) is referred to as the recognition stage, and constitutes an active enzyme system capable of modifying the next series of complement factors, enabling them, in turn, to bind to the membrane. This second or activation stage (figure 5.5b) involves an enzymic modification of the next two complement factors, C_2 and C_4, which combine to cleave a peptide from the third complement factor (C_3), all of which bind to the membrane and are competent to catalyse the final membrane-attack process. The attack complex (figure 5.5c), formed by the association of equimolar amounts of complement factors C_5 to C_8 and up to six molecules of C_9 (only three are necessary to cause cell lysis) assembles spontaneously. The $C_{\overline{56789}}$ complex alone can cause lysis of the target cell, but it is approximately three orders of magnitude more active when modified by the activated C_1 to C_4 complex attached to the membrane. This modification

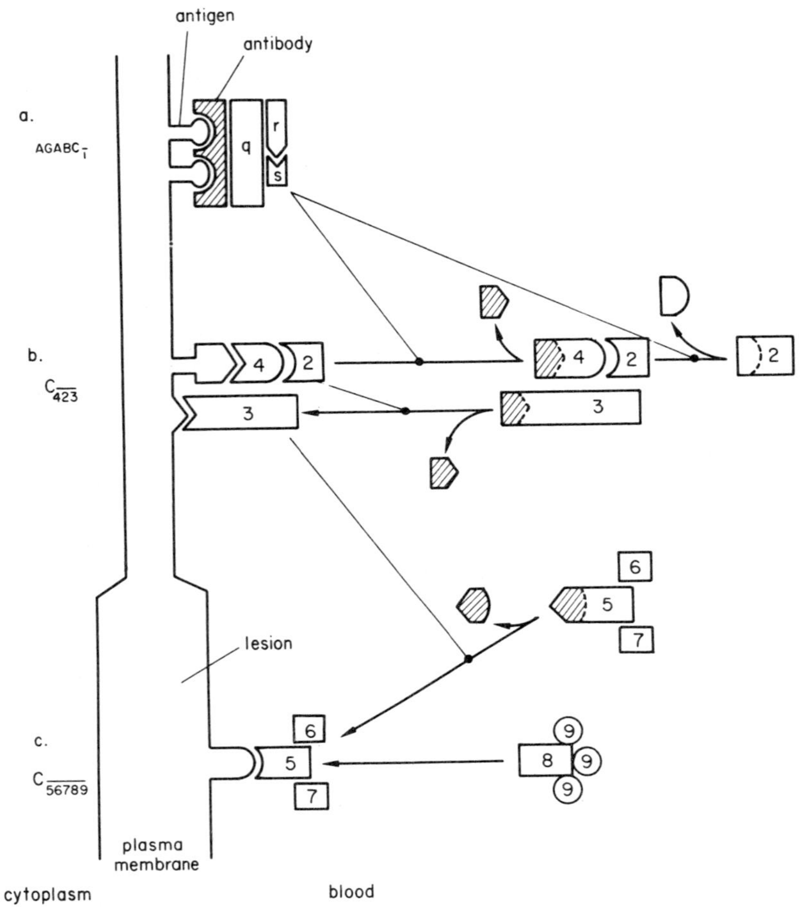

Figure 5.5 Complement fixation and lytic reactions. The three functional units of complement are illustrated.

(a) Recognition: Circulating antibody attaches specifically to cell-surface antigens (AGAB). The first complement factor binds to this and becomes activated, forming AGABC$_{\bar{1}}$.

(b) Activation: $C_{\bar{1}}$ activates both C_4 and C_2 by removing a peptide from each enabling them to combine with cell-surface features to form $C_{\overline{42}}$. This complex catalyses the second step of the activation process, the cleavage of a peptide from C_3, which binds to the membrane to give $C_{\overline{423}}$.

(c) Membrane attack: $C_{\overline{423}}$ catalyses C_5 hydrolysis and together with C_6 and C_7 binds to the membrane followed by the C_8 and C_9 components. The presence of $C_{\overline{56789}}$ causes a thickening and a lesion in the membrane through which the contents of the cell can escape (adapted from reference 21).

involves the removal of a small peptide from C_5, exposing a site capable of attaching the attack complex to the membrane, which results in rapid lysis of the target cell. The participation of such a large number of factors required to bind specifically in a cascade manner obviously provides a means of exerting a strict control over the lytic process.

The mechanism of lesion formation is not known but it is likely to be associated with a rearrangement of membrane components in the region of the assembled attack complex and may even involve some enzymic activity. Examination of the size and shape of the attack complex by electron microscopy suggests that the lytic components are arranged in a doughnut configuration of about 10 nm diameter with a hydrophilic core large enough to permit the passage of substances of high relative molecular mass out of the cell. It is still not clear, however, whether or not the pore extends completely through the membrane, since freeze-cleaved replicas prepared from membranes of lysed cells indicate that the attack complex penetrates only the outer leaflet of the membrane. Whatever the nature of the lesion, Lachmann *et al.*[22] could find no evidence that hydrolysis of phospholipids is required either to create a hole or to facilitate the rearrangement of the membrane components.

The peptides released from complement factors C_3 and C_5 during the activation process have been shown to participate in other membrane-mediated cytotoxic reactions. The peptide derived from C_3, for example, binds to the plasma membrane of target cells and promotes a chemotactic attraction of polymorphonuclear leucocytes, which then phagocytose and kill the cell. Moreover, this peptide, and also that cleaved from the C_5 complement factor, can bind to the plasma membrane of mast cells and initiate the release of histamine in an amount related directly to the number of peptide molecules bound. Serum complement factors, in addition to their pathogenic role, are also thought to be concerned in certain physiogenic reactions. The interaction of complement factors with platelet membranes, for example, is believed to mediate their coagulation during blood-clot formation. Other physiological functions of serum complement factors will no doubt come to light as research progresses in this expanding field.

5.1.10 Cell Transformation Induced Via Membrane Receptors

Certain cells can be transformed to different physiological states by the interaction of specific agents with cell-surface receptors. The T-cell lymphocyte, as we have seen, can be transformed by foreign antigens from a resting state to an active cell undergoing mitotic division. Lymphocyte transformation can also be induced experimentally using a variety of membrane-specific mitogenic agents (see figure 5.6), and such techniques have been useful in studies of the binding process and in following subsequent cellular responses. These agents include antibodies directed against lymphocytes (anti-lymphocyte sera), streptolysins and certain proteins derived from plant material referred to as agglutinins or lectins[24]. Phytohaemagglutinin from kidney bean is an example of a potent lymphocyte mitotic agent, which, when added to cells, causes them to undergo mitotic division. Within five minutes of lectin binding to the lymphocyte membrane there is an increase in phosphatidylinositol turnover followed shortly thereafter by changes associated with DNA transcription. RNA synthesis commences within two hours of the binding and, after about twenty hours, morphological transformation to a blast cell can be recognised. The incorporation of thymidine into nuclear DNA, and subsequent mitotic division ten to twenty hours later, completes the transformation. The chain of events within the cell appears to be mediated solely by the initial binding of the lectin to the receptor site although the bound antigen may subsequently enter the cell by pinocytosis of the plasma membrane.

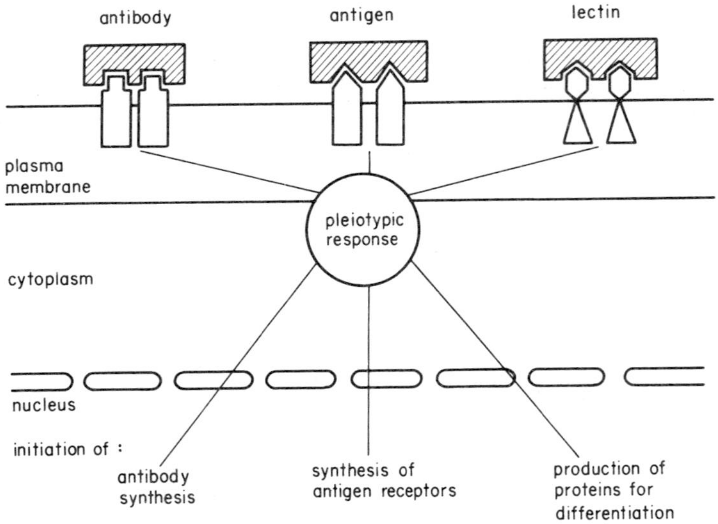

Figure 5.6 Mechansims of transformation in lymphocyte activation. Antibody, antigen and lectin bind initially to specific receptors or binding sites on the lymphocyte surface. Each agent binds to two or more receptors in a multivalent fashion. The mechanism by which the lymphocyte responds to the primary binding signal appears to be the same, irrespective of the nature of the stimulus, and involves changes in the plasma membrane, cytoplasm and nucleus in a manner resembling cells undergoing active growth. The ultimate phenotypic response, however, depends on whether the lymphocyte is differentiated into a T-cell or a B-cell, since this determines the type of proteins that are transcribed from nuclear genes. These specific responses are characteristic of the particular cell type; for example, T-cells synthesise antigen receptors and B-cells produce specific antibodies (from reference 23).

Another type of transformation is that from a normal to a malignant state[25,26]. These transformations can be induced in susceptible cells by carcinogenic compounds, radiation of various wavelengths and certain viruses. Viral transformation, illustrated in figure 5.7, involves an attachment of the virus to specific sites on the plasma membrane. The nature of these sites and of the binding process, however, is still obscure. In contrast to mitogenic transformation of lymphocytes, it is known that viral nucleic acids penetrate the plasma membrane and subsequent cellular events are almost entirely subservient to this material. One of the most notable characteristics of malignancy is that the properties of the cell surface are markedly different from normal cells. A manifestation of these changes is the failure of the cell to cease growing when a certain critical cell density is achieved (see section 3.2.3) and the cells acquire the capacity to invade normal tissues. Although changes in chemical composition and topography of the plasma membrane are known to be associated with viral transformation, the way in which viruses are able to alter the properties of the membrane is not known.

5.2 The Endoplasmic Reticulum and Golgi Complex: The Role of Membranes in Protein Synthesis

Proteins are synthesised in the cytoplasm by a complex series of reactions in which various proteins, nucleic acids and an assortment of cofactors are required.

A considerable amount of energy is needed to drive these reactions and this is supplied in the form of ATP and GTP. Mitochondria are also able to synthesise certain proteins, but the process, although similar in some respects to cytoplasmic protein synthesis, resembles in many ways the mechanism of protein synthesis by prokaryotic cells. It has been known for some time that the rate of protein

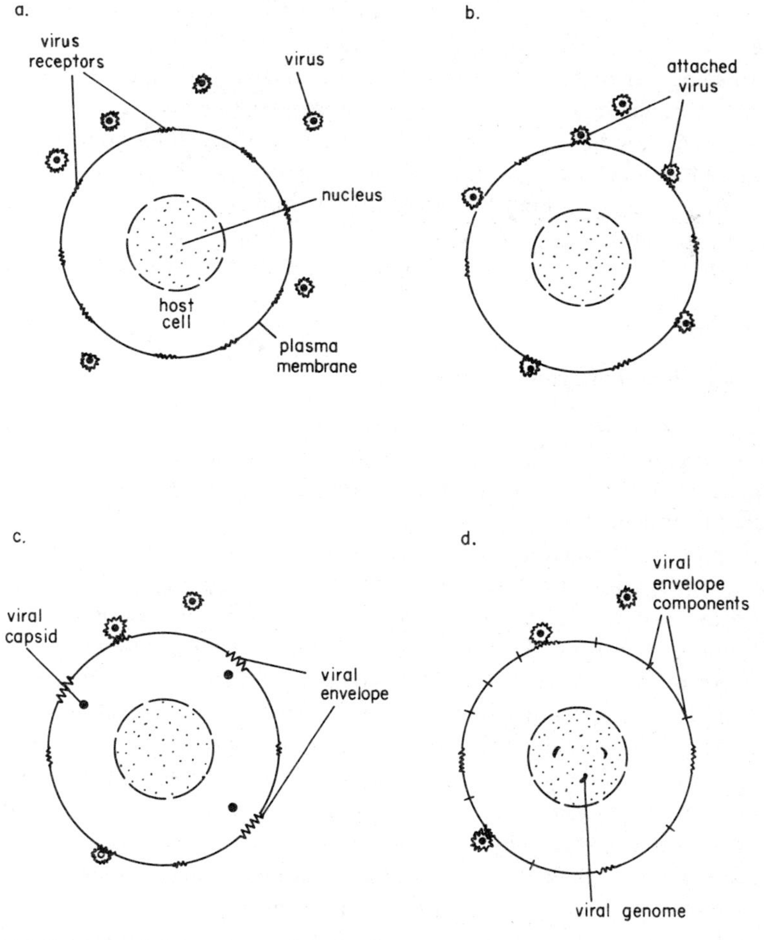

Figure 5.7 Stages in viral transformation of cells.

(a) The host cell possesses specific viral receptors on the plasma membrane to which virus particles bind.

(b) In some cases, penetration of the host plasma membrane does not follow immediately after binding of the virus.

(c) Viral capsid is released into the cytoplasm; where the virus is enveloped, the lipoprotein components of the envelope become incorporated into the plasma membrane of the host cell, and only the viral capsid enters the cytoplasm. The nuclear protein–lipid capsid is degraded by host enzymes releasing the viral genome (this may be RNA or DNA, depending on the type of virus).

(d) The viral genome migrates to the nucleus of the host and eventually may become incorporated into the host genome. This process, in the case of RNA-type virus, is accomplished by the operation of an RNA reverse transcriptase enzyme. Phenotypic changes in the host cell often accompany expression of viral genome and this includes alteration in the surface properties of the plasma membrane.

synthesis decreases drastically when the cell is disrupted. Homogenisation of most tissues, even when carefully controlled so as to minimise damage to subcellular organelles, reduces the rate of protein synthesis to only 1 or 2 per cent of that observed in intact tissue. This implies that some degree of structural integrity is required to maintain an efficient synthetic capacity, possibly by the segregation of discrete precursor pools. Not only is cellular organisation disrupted by homogenisation, but the concentration of soluble cofactors is invariably reduced, and this in itself can cause a reduction in synthetic activity. The synthesis of some proteins also appears to be more sensitive to cellular disruption than that of others. The synthesis of albumin by intact liver cells, for example, which accounts for 3 to 4 per cent of all proteins synthesised by intact cells, is not carried on by cell-free liver preparations, although they continue to synthesise other proteins at a much reduced rate. Various subcellular membranes are known to be involved in the orderly passage of albumin and other secretory proteins through the cell but the influence of this on the rate of protein synthesis is not clear.

5.2.1 Segregation of Protein-synthetic Sites

Proteins are assembled on ribosomes and, in eukaryotic cells, a certain population of ribosomes is always found associated with the endoplasmic reticulum, constituting a separate and distinctive form of this membrane, the rough-surface endoplasmic reticulum (see figure 1.4). The proportion of attached ribosomes varies from one cell to another but is usually higher in cells actively secreting protein, such as those of the liver and pancreas, where more than 80 per cent of ribosomes may be associated with the membrane[27,28]. Cells of nonsecretory tissue, on the other hand, have very few attached ribosomes; in skeletal muscle only 10 per cent, and in cerebral cortex, 20 per cent of ribosomes are membrane-bound and these proportions remain relatively constant under conditions that induce changes in ribosome distribution in secretory tissue. The significance of the free and membrane-bound ribosome populations becomes apparent when the type of proteins sysnthesised at each site are compared. The synthetic site of albumin, which is produced by liver cells and secreted into the bloodstream, has been investigated by Ikehara and Pitot[29] who have perfected an immunological technique to distinguish protein synthesis on free or membrane-bound ribosomes. They prepared radioisotopically labelled antibody fragments, which bind specifically to ribosomes synthesising albumin, and found that 85 per cent of this protein was produced on membrane-attached ribosomes. The remainder was synthesised on free ribosomes and is consistent with the fact that 15 per cent of albumin is present in the cytoplasm of the liver cell and not in membrane-enclosed cysternae. Furthermore, in studies of a minimal defective hepatoma cell line (Morris 5123), which can synthesise but not secrete albumin, it was found that the site of albumin synthesis was almost entirely on free ribosomes. Similar studies of protein-synthetic sites in other tissues (see section 5.2.4) has led to the conclusion that proteins destined for export from the cell are synthesised exclusively on the rough-surface endoplasmic reticulum.

The site of synthesis of proteins for domestic use is not so well defined and it appears they may be synthesised on either free or membrane-bound ribosomes. Serine dehydratase for example, a cytoplasmic enzyme of liver also examined by Ikehara and Pitot[29], is synthesised on both free and bound ribosomes in fasting rats; however, if the enzyme is induced by dietary means and the administration of glucagon, most of the additional synthesis is performed on attached ribosomes. Other studies with liver have indicated that soluble intracellular proteins such as ferritin and arginase are synthesised mainly on free ribosomes, whereas the membrane protein NADP–cytochrome C reductase, found in the endoplasmic reticulum, appears to be synthesised on both free and membrane-bound ribosomes. Additional evidence supporting the contention that not all proteins synthesised on attached ribosomes are secreted from the cell has come from studies of the human malignant HeLa cell line. It has been estimated that in these cells about 15 per cent of the total ribosome population is membrane-bound yet only 2 per cent of newly synthesised protein is ultimately exported from the cell even though the rate of protein synthesis is probably greater on attached ribosomes[30]. Other workers believe that proteins required for cell growth, differentiation and proliferation are synthesised predominantly on membrane-associated ribosomes.

5.2.2 Attachment of Ribosomes to the Endoplasmic Reticulum

Free ribosomes in the cytoplasm of eukaryotic cells exist in discrete exchangeable pools of native 40S and 60S subunits, 80S monoribosomes and polyribosomes in which a number of 80S ribosomal units are attached to a single mRNA template. If, as seems likely, specific proteins are synthesised exclusively on membrane-associated ribosomes whereas others are manufactured predominantly on free ribosomes, then there must be a mechanism for selecting the site of synthesis of particular proteins. How this discrimination is achieved is not known, but since mRNA codes for specific proteins it might be expected that this component will be involved in some way. Membrane-associated ribosomes may be assembled by attachment of a 40S ribosome-mRNA–tRNA complex to a 60S ribosomal subunit bound to the endoplasmic reticulum. An alternative mechanism whereby the complete ribosome assembly is attached to the endoplasmic reticulum by interaction with the nascent or growing peptide chain has been considered by Rolleston[31]. The work of Shires *et al.*[32] tends to eliminate the existence of two separate populations of ribosomes each synthesising a separate group of proteins. They found that virtually all cytoplasmic polyribosomes are capable of binding to membranes from which all attached ribosomes had been removed by prior treatment with ribonuclease, and these bound with about the same affinity as ribosomes originally associated with the membrane. Free polyribosomes were also able to bind to smooth-surface endoplasmic reticulum and other membranes, but the association in most instances was not specific, since the nascent peptide did not transverse the membrane (see section 5.2.4). There appears to be a finite number of discrete ribosome-binding sites on the rough-surface membrane to which ribosomes can attach specifically, and the number of sites is thought to be the main factor in determining the proportion of free and membrane-bound ribosomes.

There have been a number of attempts to characterise the ribosome binding site on the rough-surface endoplasmic reticulum. The functional 80S ribosomes are known to be attached to the membrane by interaction through the larger (60S) ribosomal subunit, and although the 40S subunit can also bind to the membrane it does so with much less affinity than the 60S subunit. Electrostatic forces appear to be involved in the binding because a certain proportion of membrane-bound ribosomes can be removed by treatment of rough-surface endoplasmic reticulum with high salt concentration (1 M) combined with calcium sequestering agents. Moreover, membranes treated with ribonuclease will not bind ribosomes in the presence of high salt concentrations. Salt treatment, however, denatures ribosomes and renders them functionally inactive, and it is likely that other membrane components are extracted along with the ribosomes even though salt-extracted membranes appear to remain intact. Ribosomes can be detached in a functional state with detergents but membrane integrity is disrupted by detergent treatment. Recent experiments conducted by Adelman *et al.*[33] suggest that other forces stabilise the attachment of ribosomes to the membrane (see figure 5.8). They found that about 40 per cent of membrane-bound ribosomes in liver could be detached by treatment with moderately high salt concentrations (0.75M KCl), and these were characterised by the absence of very short nascent peptides suggesting that this category of ribosomes bind to the membrane predominantly by electrostatic interactions (figure 5.8a and b). Nearly all the attached ribosomes, however, could be released by salt treatment if puromycin was added beforehand. This agent is an analogue of the aminoacyl-adenosine terminus of aminoacyl-tRNA and blocks polypeptide elongation by terminating the peptide sequence. The nascent puromycin-peptide derivative remains associated with the membrane fraction after the ribosome has been released (figure 5.8d). It was suggested that the tighter binding of ribosomes with long nascent peptides is probably due to stabilisation of the primary electrostatic interaction of the ribosome with the membrane, either by a direct interaction of the nascent peptide with the membrane or by the development of protein tertiary structure on the distal side of the membrane. Considerable folding of the polypeptide chain would be expected, considering that nascent peptides may attain linear dimensions of more than 0.1 μm before the sequence is completed.

The interaction of ribosomes with the endoplasmic reticulum is unlikely to involve enzymic activity because binding is unaffected by temperature. Furthermore, a preliminary salt extraction of the membrane does not impede subsequent ribosome-binding, but proteolytic enzyme digestion completely abolishes binding capacity, suggesting that an intrinsic membrane protein is required for ribosome attachment. An indication that this protein may interact with steroid hormones to constitute a ribosome-binding site has come from the work of Rabin and his colleagues[34,35]. They found that steroid hormones promote the binding of soluble ribosomes to the smooth-surface endoplasmic reticulum in a sex-specific manner. Thus oestradiol causes ribosome attachment to membranes from male rat liver but not to the corresponding female membranes, whereas testosterone promotes ribosome binding to female rat-liver membranes only. The ribosome-binding characteristics were directly correlated with the number of specific hormone-binding sites on the respective membranes.

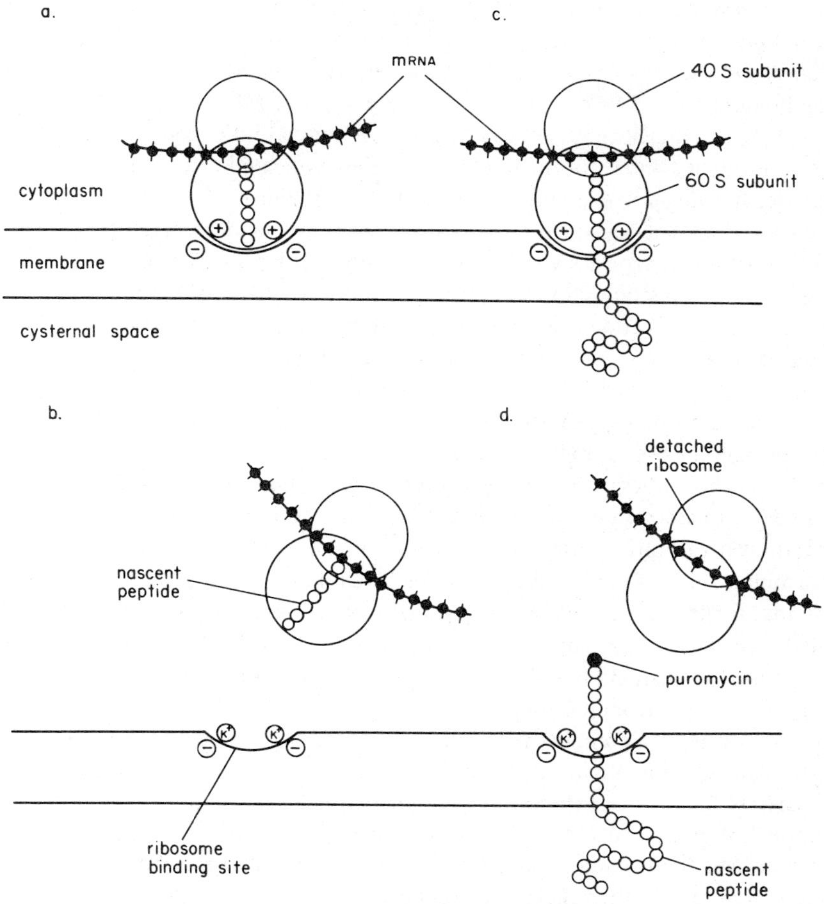

Figure 5.8 Characteristics of ribosome attachment to the endoplasmic reticulum. The complete functional ribosome is attached to the membrane via the 60S subunit (a) and can be detached with salt (0.75M) provided the nascent peptide is short (b). As the nascent peptide becomes longer and penetrates across the membrane and into the cysternal space (c), the ribosome can no longer be detached with salt unless the sequence is terminated beforehand with puromycin (d) (see reference 33).

5.2.3 Amino Acid Activation

Lipids have been implicated in the activation of amino acids prior to their incorporation into proteins and some of the reactions are thought to be performed at membrane sites. Amino acid activation and nucleotide coding reactions are catalysed by aminoacyl-tRNA synthetase enzymes specific for each amino acid. The steps in the reaction have been described in detail by Loftfield[36] and involve the formation of an aminoacyl derivative of AMP in which the energy of the pyrophosphate bond of ATP is conserved in the formation of the acyl linkage. This bond is subsequently transferred to the specific tRNA coding for the particular amino acid. Aminoacyl-tRNA synthetases have been purified from liver by Bandyopadhyay and Deutscher[37], who showed that all the acitivity could be recovered from a post-membrane centrifugation pellet in the form of a single enzyme complex containing 20 to 25 per cent lipid. The lipid consisted almost entirely of cholesterol esters, which are not normal constituents of cell membranes, and the major component was cholesterol-14-methylhexadecanoic acid. The lipids appear to be concerned with preserving the association of the complex in an active form, since its removal leads to a dissociation of the complex, but it is also thought to assist in the binding of the enzyme complex to ribosomes. Hampel and Enger[38] examined the subcellular distribution of all twenty aminoacyl-tRNA synthetases in cultured hamster ovary cells and observed that a large number of these enzymes were associated with membrane fractions. They concluded, from studies involving tissue disruption by nitrogen cavitation procedures and treatment with nonionic detergent, that the enzymes were likely to be associated with cell membranes in intact tissues.

Certain hormones including insulin, thyroxine and growth hormone are able to stimulate protein synthesis in responsive tissues. The amino acids incorporated into protein in some stimulated tissues are derived preferentially from exogenous sources without these amino acids equilibrating with the intracellular pool. This has been explained by some workers as being due to the formation of amino acid derivatives during entry into the cell, which distinguishes them from the free amino acids of the intracellular pool. The binding of amino acids directly to aminoacyl-tRNA synthetase at the plasma membrane seems unlikely because the calculated rate of diffusion of such a complex through the cytoplasm would be insufficient to sustain observed rates of protein synthesis in most cells. An alternative mechanism has been proposed by Hendler[39], who has suggested that amino acids form a complex with lipids during passage through the plasma membrane and migrate in this form through the cytoplasm to aminoacyl-tRNA synthetase located near the site of protein synthesis. He supports this view by demonstrating that such complexes can be extracted from tissues with organic solvents, and that these so-called lipoamino acids are incorporated into protein much more readily than their water-soluble counterparts. It could be argued that entry of lipoamino acids into the cell is favoured compared with the more polar free amino acids; similar results, however, were obtained in cell-free systems, where such advantages are superfluous. All amino acids are capable of forming complexes with lipids with both the amino and carboxyl groups participating in the binding, and free amino acids can be released from the complex by hydrolysis. It should be emphasised that the role of membranes and lipids in amino acid activation is mainly speculative at this stage and many aspects of this problem represent fertile ground yet to be explored.

5.2.4 Synthesis of Proteins Destined for Secretion from Pancreatic Exocrine Cells

Many cells synthesise proteins that are eventually secreted from the cell. These proteins, some of which are hormones and enzymes potentially harmful to the cell producing them, are synthesised almost exclusively on membrane-associated ribosomes. The action of secretory products on the cell is prevented in two ways; firstly they are segregated from the cytoplasmic compartment by vectoral synthesis across the rough-surface endoplasmic reticulum into membrane-enclosed cysternae, and secondly they are usually synthesised as inactive precursors or zymogens and are not converted to an active form until just prior to or soon after release from the cell. The role of intracellular membranes, and particularly the function of the Golgi complex, has been established by observing the procession of specific proteins from the site of synthesis to their final secretion from the cell. The pioneering studies, conducted by Palade and his co-workers and reviewed recently by Jamieson[40], employed visible-light and electron microscopy combined with autoradiography of pancreatic exocrine cells synthesising α-chymotrypsinogen and α-amylase. The original experiments conducted by Caro and Palade[41] consisted of administering radioisotopically labelled amino acids to starved guinea pigs one hour after they had been re-fed to stimulate protein synthesis. The pancreas was removed from animals at intervals after the initial amino acid pulse, tissue sections were prepared for electron microscopy, and protein which had incorporated radioisotopically labelled amino acid was localised within the ultrastructure of the pancreatic cell by standard autoradiographic techniques. Similar studies were performed on slices of guinea pig pancreas with essentially the same results. Additional experiments were performed in conjunction with the electron microscopy to ascertain the subcellular distribution of radioisotopically labelled proteins[42,43]. It was possible, on the basis of these experiments, to divide the overall process into five separate stages (see figure 5.9).

(1) Synthesis of the polypeptide on membrane-associated ribosomes. The growth of the nascent peptide is directed through the membrane and, on termination of the amino acid sequence, the protein is released into the cysternal space.
(2) The newly synthesised protein accumulates in cysternae of rough-surface endoplasmic reticulum.
(3) The protein then percolates through the cysternal space, passing through a region where the membrane undergoes a transition from rough surface to smooth surface. This stage is completed when the protein is enclosed in smooth-surface membrane in the region of the Golgi complex.
(4) The secretory product is concentrated in vacuoles, which are thought to be derived in part from Golgi membranes. The process from synthesis up to stage (4) is completed in about forty-five minutes.
(5) This stage is essentially a storage phase in which the secretory granule remains in the cytoplasm until it is eventually released, usually in response to a specific signal originating from outside the cell. The contents of the granule are liberated to the exterior by fusion of the granule membrane with the plasma membrane.

In the pancreatic exocrine cell, synthesis of protein is performed mainly in the basal region of the cell and the product moves through several different transport phases, each involving passage within a membrane-bounded subcellular compartment, towards the apical cell region. The protein is released into the lumen, which is formed from the apical plasma membranes of a number of pancreatic cells. This directional movement of proteins through the cell is a common feature of all secretory tissue, but in certain cases, such as thyroid, movement can occur in both directions.

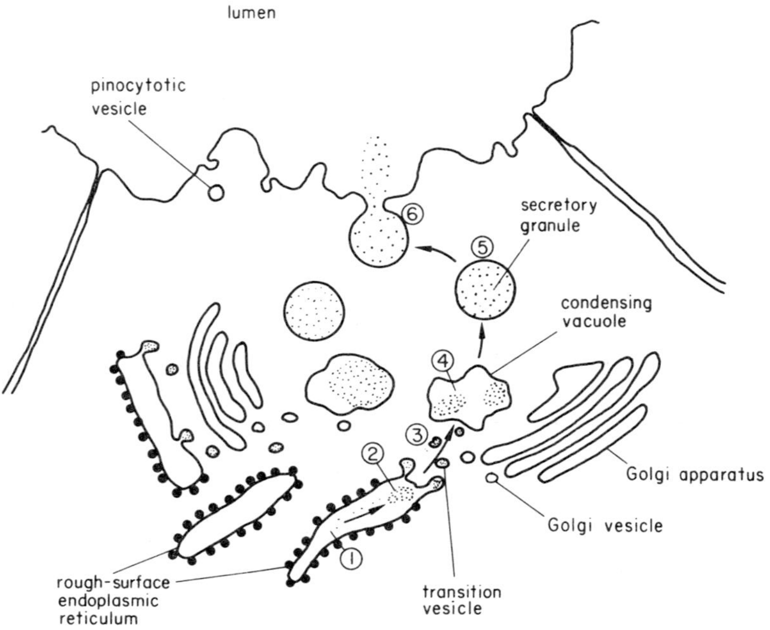

Figure 5.9 Schematic representation of the stages in the synthesis, transport and secretion of proteins from the pancreatic exocrine cell. Proteins destined for secretion are synthesised on rough-surface endoplasmic reticulum (1) and accumulate in the cysternal space of these membranes (2). Small smooth-surface transition vesicles containing secretory product bud off from the endoplasmic reticulum (3) and fuse to form a condensing vacuole (4). It is not clear whether Golgi vesicles fuse with these transition vesicles or directly with the condensing vacuole, but elements of the Golgi membranes are known to be located in the membrane of the secretory granule. The concentrated product is stored in the secretory granule (5) and the contents are eventually released to the lumen by fusion of secretory granule membrane with the apical plasma membrane of the cell (6). The direction of movement of secretory protein is indicated by arrows (see reference 40).

5.2.5 Synthesis and Storage of Secretory Protein by Other Cells

Studies similar to those of Palade's group on pancreatic zymogen synthesis have been extended to other secretory cells synthesising a variety of proteins. These include casein synthesis by mammary tissue[44] and insulin production by pancreatic β cells[45] and in both these systems the mechanism of synthesis and transport of proteins closely resembles the original observations on pancreatic α cells. As well as largely confirming the earlier work, these studies have helped to identify the essential functions of subcellular membranes in secretion, and particularly in characterising the special role of the Golgi complex. The synthesis of insulin[46,47] provides an excellent example of Golgi function, in addition to illustrating a mechanism for protecting the cell from hormonal effects of the secretory product.

Insulin is synthesised as a single polypeptide percursor, proinsulin, on the rough-surface endoplasmic reticulum of the pancreatic β cell. Proinsulin itself has only slight insulin-like effects when tested on cells normally responsive to insulin[48]. During stage (2) of the process (see figure 5.9), groups of six proinsulin peptides are assembled about two atoms of zinc by co-ordination through histidine residues of the polypeptide chain. The hexamer arrangement brings about a burial of the active insulin region of the polypeptide chain in the interior of the complex, leaving the connecting peptides exposed on the surface, where they are readily available for cleavage at a later stage. The connecting peptide serves to orientate the ends of the polypeptide chain into suitable configuration for disulphide bonding between what ultimately become the A and B chains of active insulin, and it confers low insulin-like activity by masking the determinant amino acids of these chains. Proinsulin hexamers enter the Golgi phase about ten to fifteen minutes after synthesis, and at this stage removal of the connecting peptide commences. Cleavage of the peptide involves a two-step hydrolysis by separate proteolytic enzymes, one of which has trypsin-like activity and the other is similar in action to carboxypeptidase. Both enzymes are thought to be either components of the Golgi and secretory granule membranes or confined within vesicles formed by these membranes, because insulin and connecting peptide are the only products released from the cell[49]. After cleavage of the connecting peptide is complete (the process has a half time of slightly less than one hour), the resulting insulin hexamers aggregate into crystalline structures preparatory to secretion from the cell. Insulin is released from the cell usually between six and eight hours after synthesis together with equivalent amounts of the connecting peptide, which remains inside the secretory granule after cleavage.

The mode of protein synthesis and pathway of secretion of proteins from the exocrine and endocrine (β-cells) pancreas appear to be typical of most secretory cells. The apparent differences between various cell types is mainly in the length of time that secretory product is stored in cytoplasmic storage granules. Secretory cells of the pancreas, for example, possess considerable reserves of zymogen and insulin, which can be released in appropriate amounts when required, whereas other secretory cells such as liver, which secretes albumin, and B-cell lymphocytes, which secrete immunoglobulin, have little or no reserves of these proteins in cytoplasmic storage granules. Protein is secreted from these cells at a rate that is directly related to its synthesis on the rough-surface endoplasmic reticulum. Nevertheless the method of synthesis and transport through the cell appears to be the same as in pancreatic cells, at least up to the Golgi phase.

A unique method of hormone storage is found in the thyroid gland[50] and the essential features of this process are illustrated in figure 5.10. Thyroid cells are arranged into a follicle, and the hormone precursor thyrotrophin, which is synthesised on the rough-surface endoplasmic reticulum, is transported to the lumen of the follicle, where it is stored in the form of a colloidal suspension. When the gland is stimulated by pituitary thyrotrophin, small droplets of this colloid material are taken up from the lumen by a process of endocytosis at the apical plasma membrane of the follicular cell. Membrane-enclosed colloid droplets then fuse with lysosomes in the cytoplasm, whereupon the thyroglobulin undergoes hydrolysis by lysosomal enzymes, and finally the thyroid hormones are liberated into the blood by fusion of the vesicle with the basal cell membrane. It is not known whether Golgi membranes are involved in the iodination of thyroglobulin, but clearly the Golgi is not concerned in the liberation of thryoxin from thyroglobulin, which is performed exlusively by lysosomal enzymes.

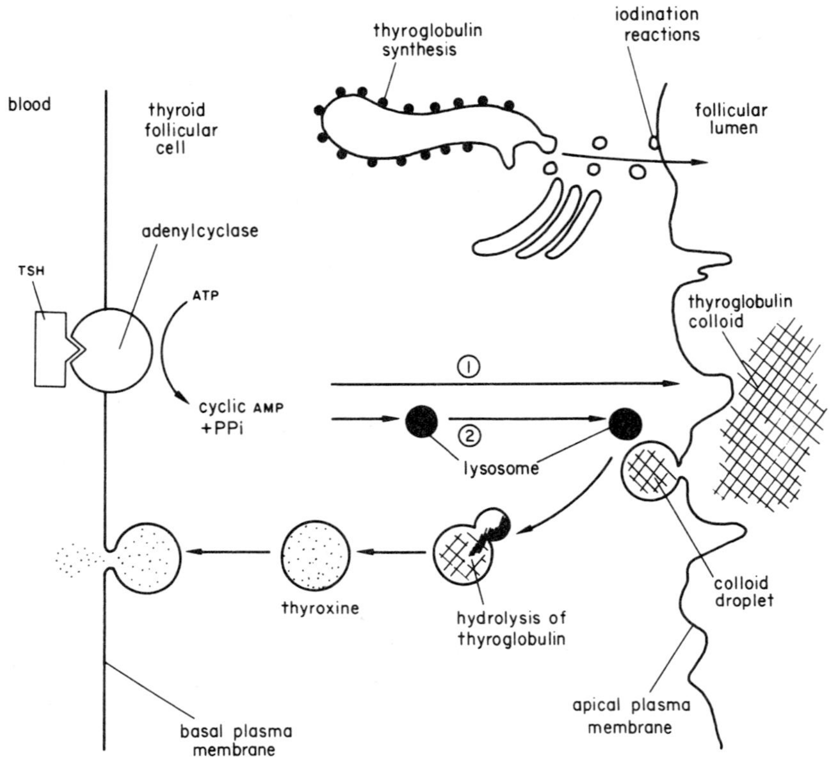

Figure 5.10 Synthesis, storage and secretion of thyroid hormone. Thyroxin is synthesised as a storage precursor, thyroglobulin, on ribosomes of the rough-surface endoplasmic reticulum and transported to the apical plasma membrane in transition vesicles (perhaps fusing with Golgi vesicles). Tyrosyl residues of thyroglobulin are iodinated at this point and the iodinated product is released, by a process of exocytosis, to the follicular lumen, where it is stored in the form of a colloidal suspension. The release of thyroxin to the blood stream is triggered by activation of adenylcyclase located in the basal plasma membrane by circulating pituitary thyrophin (thyroid-stimulating hormone, T.S.H.). Cyclic AMP is believed to activate the apical plasma membrane either directly, as shown by pathway (1) or indirectly (2) by promoting labilisation and migration of lysosomes to the apical membrane. Colloid is taken up from the lumen by endocytosis at the apical plasma membrane and droplets of colloid enclosed in this membrane enter the cytoplasm. Lysosomes fuse with colloid droplet membranes, releasing proteolytic enzymes, which hydrolyse thyroglobulin and liberate thyroxine (T_4 and T_3). Thyroid hormone is secreted into the blood by fusion of the combined vesicle membrane with the basal plasma membrane of the thyroid cell (see reference 51).

5.2.6 Membrane Interactions During Secretion

Subcellular membranes undergo changes in composition and position within the cell in connection with the movement and processing of the secretory proteins. Jamieson and Palade[52] have shown that these membrane changes do not depend on continuing protein synthesis, since secretory products can be transported out of the cell from the site of synthesis even when synthesis of new protein is prevented by specific blocking agents. Proteins required for the synthesis of new membranes must therefore be supplied either from a cytoplasmic pool of membrane proteins or by re-utilisation of existing membrane components. Two situations are envisaged in which membrane transformation and re-utilisation are likely to be involved in the transport process, and these are shown schematically in figure 5.11. The first involves the passage of transition vesicles between the rough-surface endoplasmic reticulum and the periphery of the Golgi complex. Condensing vacuoles are formed by coalescence of large numbers of these small vesicles, but the amount of membrane supplied exceeds that required to envelop the concentrated product. It has been suggested that either small vesicles or dissociated membrane components return to form new transition elements of the endoplasmic reticulum[53]. The second process concerns the mechanism whereby the surface area of the plasma membrane is maintained. The fusion of storage-granule membranes with plasma membranes during exocytosis causes an expansion in the surface area of the cell, and some method of compensating for this increase in area must operate to maintain cell size within certain limits. This could be by removal of individual components from the plasma membrane or, more likely, the formation of pinocytotic vesicles pinched off from the plasma membrane, which could then be used to replenish membranes of the Golgi complex. The transformation of membranes between the endoplasmic reticulum and the Golgi complex and between the secretory granule and the plasma membrane requires an expenditure of energy, since the transport of secretory product can be arrested at either point if the supply of ATP is blocked by the addition of respiratory inhibitors[54].

During differentiation, individual membranes must undergo certain changes

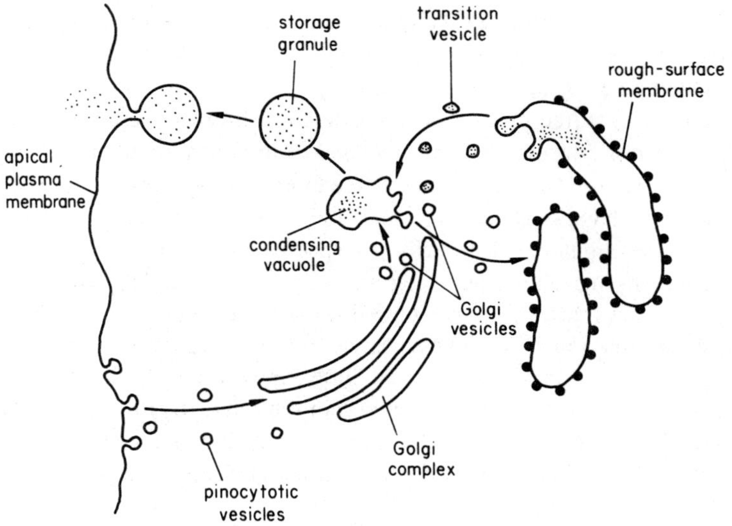

Figure 5.11 Membrane transpositions during secretion. Two suggested mechanisms of membrane transposition centred on the condensing vacuole are shown. The condensing vacuole is formed by fusion of transition vesicles from the rough-surface endoplasmic reticulum and Golgi vesicles. Vesicles budding off from the condensing vacuole may return to the endoplasmic reticulum. The condensing vacuole eventually forms a storage granule, and the membrane becomes incorporated into the apical plasma membrane during the secretory process. Elements of the plasma membrane in the form of pinocytotic vesicles may be used to replenish membranes of the Golgi complex. The cyclic nature of these membrane movements may not be as direct as indicated, because pinocytotic vesicles and vesicles derived from the condensing vacuole could also fuse with other sub-cellular organelles such as lysosomes.

that reflect their functional transformation and this probably involves a considerable rearrangement of the existing constituents, possibly in association with deletion and/or addition of new components to the membrane. These changes in the membrane may be essential for directing the fusion of specific membranes at the appropriate time and place, thus ensuring overall continuity of the secretory process. Alterations to the membrane are thought to be performed primarily during the Golgi phase[55], but have been known to extend into the storage phase as well. Biochemical studies have demonstrated the presence of multi-enzyme glycosylation systems in Golgi membranes. These enzymes are known to mediate in the attachment of specific sugar residues or complex oligosaccharides on to peptide precursors of membrane glycoproteins as well as to secretory proteins confined within the vesicle. Certain polysaccharides incorporated into membrane components participate in cell-surface phenomena following fusion of the vesicle with the plasma membrane and subsequent exposure on the cell surface (see section 5.1). It has been suggested that Golgi vesicles may also fuse with other sub-cellular organelles such as lysosomes, where specialised components of the vesicle membrane may serve as informational messenger agents. However, there is no direct evidence to support such a function. Proteins may also be phosphorylated or sulphated by specific phosphokinases and sulphokinases, which have also been found associated with Golgi membranes of some cells. It is thought that these enzymes may be concerned primarily with modifying secretory proteins rather than proteins of the vesicle membrane. In addition to biochemical changes, morphological alterations have also been observed, which are consistent with transformation of membrane from endoplasmic reticulum to plasma membrane during the Golgi phase.

The release of stored products from different cell types is initiated by diverse, yet highly specific, agents. Insulin is released from the endocrine pancreas, for example, in response to an increase in circulating blood-glucose concentration, whereas catecholamine release from the adrenal medulla or sympathetic nerve endings can be stimulated electrically. Irrespective of the mechanism of stimulation, the underlying cellular responses are believed to be similar. This is partly based on the fact that calcium is required for the release of all secretory products, and stimulation is usually associated with an elevation of intracellular cyclic AMP concentration. Rasmussen[56] has proposed that cyclic AMP and calcium serve as secondary messengers within the cell, acting in concert to alter the interaction between subcellular components, including the storage granule and the plasma membrane, and leading to fusion and release of products.

5.3 Mitochondrial Membranes–Compartmentation and Energy Transduction

In the preceding discussion dealing with the secretion of cellular products, organisation of the various membranes into operationally separate and distinctive subcellular compartments was found to play an important role in directing the synthesis and transfer of secretory protein from one site to another within the cell. On another level, subcellular compartments are also concerned with the regulation of metabolic reactions, and examples of the relationship between different metabolic compartments and substrate pools in the liver cell have been cited recently by Gumaa *et al.*[57]. The mitochondrion is perhaps one of the most striking examples of metabolic compartmentation. In addition to segregating metabolite pools, however, the mitochondrial membranes also perform other highly specialised functions concerned with the transformation of energy derived from metabolic reactions. The free energy derived from the oxidation of organic molecules is conserved into ATP, the biologically useful form of energy for the cell, largely by reactions associated with mitochondria. The high-energy terminal pyrophosphate bond of ATP is used in turn to drive synthetic reactions or to perform various categories of cellular work. Morphologically, mitochondria possess two well-defined compartments, an outer inter-membrane space bounded by the inner and outer membranes, and an inner matrix, which is in direct contact only with the inner membrane (see figure 1.6). A region of the inter-membrane space between cristae membranes is sometimes referred to separately as the intra-cristal space, but any functional distinction between these two regions is not clear. The metabolic pathways by which substrate molecules are catabolised appear to be segregated between these two compartments mainly as a consequence of the highly selective permeability of the inner mitochondrial membrane. The other unique feature of this membrane is that it contains all the enzymes and cytochromes of the electron-transport chain required to oxidise reduced substrates. These respiratory catalysts, which form integral components of the membrane, are organised in such a way that the oxidative energy is conserved and transformed in phosphorylation reactions with the aid of associated coupling factors.

5.3.1 Structural and Functional Approaches to Defining Mitochondrial Compartments

There are two experimental approaches by which a topographical relationship can be established between various mitochondrial functions and the compartments defined by the mitochondrial membranes. The first is a structural approach, which seeks to assign a functional role to each of the mitochondrial compartments and depends on locating enzymes and cofactors required to perform specified metabolic reactions. Histochemical methods in conjunction with electron microscopy have been used for this purpose but, owing to the complex folding of the cristae membranes, the exact location of certain membrane-bound enzymes is difficult to establish unequivocably. Seligman *et al.*[58], however, have used histochemical techniques successfully to show that cytochrome oxidase (cytochrome $a + a_3$) activity located in the inner mitochondrial membrane culminates in the oxidation of cytochrome c on the outer surface of this membrane. The main impetus to the structural approach has resulted from the development of satisfactory methods for separating the inner from the outer membrane. Essentially this is done either by disrupting the outer membrane by treatment with bile acid detergents such as deoxycholic acid (the outer membrane is preferentially disrupted by bile acids because of its relatively high cholesterol content) or by osmotic lysis in the presence of phosphate and other ions. In both procedures the inner membrane appears to remain intact and contains a full complement of matrix constituents. Analysis of enzyme activities associated with the matrix and inner mitochondrial membrane can then be compared with those associated with the outer membrane or recovered in soluble form after removal of the outer membrane. In addition, the characteristics of the various ion and metabolite transporting systems of the inner membrane can be examined directly in these submitochondrial particles.

Since the inner membrane is not disrupted by bile acids or osmotic lysis, the surfaces of the vesicle formed after removing the outer membrane face the same relative direction they did in the intact mitochondrion. A reversal of polarity can be achieved by treating the vesicles with phospholipase C or ultrasound, in which case the membrane particles originally facing the matrix or inner surface (see section 1.1.5) are now observed on the outside of the vesicles. The reversal of polarity has been elegantly demonstrated by Racker *et al.*[59], who prepared thin sections of mitochondria to which electron-opaque ferritin particles had been added before and after phospholipase C treatment. They observed small vesicles, presumably formed from the inner membrane, packaged within an apparently intact outer membrane; however, ferritin particles were trapped inside the vesicles only if the ferritin was added before phospholipase treatment. This suggests that ferritin cannot penetrate the inner membrane even after phospholipase treatment, so its presence inside the vesicles can only be explained if the contents of the vesicle originate from material derived from the inter-membrane space. A functional reversal of submitochondrial particles prepared by ultrasonic treatment has also been demonstrated, since these vesicles are capable of oxidising exogenous but not trapped pyridine nucleotides, a feat that cannot be performed by osmotically derived inner-membrane preparations. Submitochondrial particles of inner membrane retain certain functional features, including that of selective permeability, so that comparative studies with intact inner-membrane vesicles have been extremely useful in studies concerned with the vectorial transport of ions and metabolites across the inner membrane.

The second approach to defining metabolic compartments is based on mitochondrial function alone. This method consists of monitoring the behavioural properties of intact mitochondria when they are supplied with certain substrates and cofactors and equating these with the existence of separate compartments or differences in permeability between one or other of the mitochondrial membranes. Harris and Manger[60] found evidence for compartmentation when they examined respiration rates of substrates that were added either individually or in selected combinations to intact mitochondria. They found, for example, that oxidation of β-hydroxybutyrate was additive in combination with various tricarboxylic-acid-cycle intermediates, with the exception of succinate, where there was evidence of competition presumably because succinate accumulation by mitochondria was inhibited by β-hydroxybutyrate. This and other evidence indicated that β-hydroxybutyrate was confined in a compartment separate from the other metabolites, many of which competed against each other for transport into the mitochondria (see section 5.3.2). Some combinations of tricarboxylic-acid-cycle intermediates, however, provided a greater rate of oxidation than the sum of the separate rates, whereas with others it was less. They interpreted enhancement of oxidation rate as a removal of inhibitory products, which might compensate for any mutual competition between substrates, in contrast to lower oxidation rates where substrates are mutually competitive. Stimulation of malate or succinate oxidation by adding glutamate, for example, is believed to result from the removal of inhibitory oxaloacetate by transamination. In addition to metabolites, the accessibility of mitochondrial compartments to ions and other substances has been measured by their effect on the functional activities of mitochondria or by their osmotic properties, which can be observed conveniently by changes in light-scattering properties of mitochondrial suspensions as they swell and shrink. Notwithstanding the difficulties outlined above, the functional approach has proved useful in establishing the site of reactions that depend on an intact structure and in particular those that take place in the inter-membrane space.

In general, anaerobic pathways are located in the inter-membrane space or extramitochondrially, while oxidations are performed either in the matrix or in the inner mitochondrial membrane. The anaerobic catabolism of complex substrates such as the conversion of carbohydrates to pyruvate and of amino acids to α-keto acids takes place mainly in the outer compartment, where the appropriate enzymes for these reactions are found. Some of these enzymes are freely soluble in the aqueous region while others are bound to either the inner or the outer mitochondrial membranes. They are, however, accessible to substrates within the inter-membrane space. The matrix compartment contains all the enzymes of the tricarboxylic acid cycle as well as those concerned with the oxidation of fatty acyl CoA derivatives (β oxidation).

5.3.2 Permeability and Substrate Transport Across Mitochondrial Membranes

The permeability characteristics of the inner and outer mitochondrial membranes are found to be markedly different. The outer membrane is freely permeable to both charged and uncharged molecules with a relative molecular mass of less than 10 000. This is sufficient to retain most soluble enzymes within the inter-membrane space and yet allow substrates and cofactors to pass freely between this region of the mitochondria and the cytoplasm. The inner membrane, on the other hand, is impermeable to large molecules, and only certain small uncharged molecules like water, oxygen and carbon dioxide, all with relative molecular masses of less than 150, are freely permeative. Urea and glycerol are other compounds that have unimpeded access to the mitochondrial matrix, whereas sucrose is only able to enter the inter-membrane compartment of intact mitochondria. Chappell *et al.*[61] have described several methods that can be used to study the permeativity of mitochondrial metabolites through the inner mitochondrial membrane. Taking the fact that this membrane is freely permeable to water, the transport of substrates can be observed by osmotic swelling and contraction of the mitochondria. Since most substrates are anions, penetration of the membrane must be accompanied by either an equivalent cationic charge or a corresponding exit of anions in order to retain electrical neutrality. It is customary therefore to add a freely permeative cation such as ammonium or methylammonium, neither of which interferes with energy-linked transport processes. Mitochondria swell when suspended in iso-osmotic ammonium salts of permeant anions at a rate related to the amount of anion accumulated in the matrix. When a suspension of mitochondria swells, there is a decrease in apparent absorbency of light according to the relationship $V \propto 1/A$, where V is the matrix volume and A is the apparent absorbence, so changes in V can be observed directly. An alternative procedure is to add a small amount of a concentrated ammonium salt to mitochondria suspended in iso-osmotic ammonium chloride and, after an initial contraction to compensate for addition of the anion, the mitochondria swell as the anion penetrates. No subsequent swelling is observed when the anion is unable to penetrate the membrane. An isotopic method, the so-called *space technique,* has been used to determine penetration of substrates into the inter-membrane space of the matrix. In this method, substrate-depleted mitochondria are equilibrated with ^3H_2O and ^{14}C-sucrose (sucrose does not enter the matrix of intact mitochondria) before addition of the test compound. At appropriate intervals, mitochondria are sedimented and the substrate concentration is measured in the mitochondria and supernatant. Substrate or respiratory inhibitors must also be added in these experiments to prevent metabolism of the compound under study. These techniques have been used to demonstrate that succinate and α-oxoglutarate penetrate only the sucrose-accessible space of mitochondria, but if malate is present both are able to enter the matrix (^3H_2O-accessible space) as well. The final procedure described by Chappell *et al.*[61] was devised in an attempt to overcome objections to the methods outlined above in that they are not conducted under strictly physiological conditions. In this method the metabolism of substrate anions as they penetrate the matrix is observed by measuring changes in the redox state of endogenous pyridine nucleotides. For these experiments to be successful it is necessary firstly, to convert all the pyridine nucleotides to their oxidised form by adding agents to uncouple oxidative phosphorylation, and secondly to block reoxidation during oxidation of the substrate anion with the aid of respiratory chain-blocking agents. The rates at which pyridine nucleotides are reduced can be measured spectrophotometrically or fluorometrically, and this is related to the rate of entry of the substrate into the matrix compartment.

The general conclusions derived from these studies can be summarised as follows: small monocarboxylic acid anions like pyruvate, acetoacetate and β-hydroxybutyrate, which are generated outside the mitochondria or in the inter-membrane space, are able to move across the inner membrane, but the rate of diffusion indicates that only molecules in an undissociated form are free to do so. Dicarboxylic and tricarboxylic acids are unable to diffuse through the membrane and they must be transported by specific carrier proteins located at sites in the membrane. The transport of these acids into the matrix is coupled to the movement of anionic molecules in the opposite direction in a process known as exchange diffusion. The various transporter systems participating in substrate-exchange diffusion are illustrated in figure 5.12. It can be seen that the exchange process operates in such a way that the distribution of charge across the membrane is not disturbed. Thus, inorganic ions such as phosphate are transported into the matrix probably in exchange for hydroxyl ions on a specific carrier protein located in the inner membrane. The transport is sensitive to sulphydryl reactants

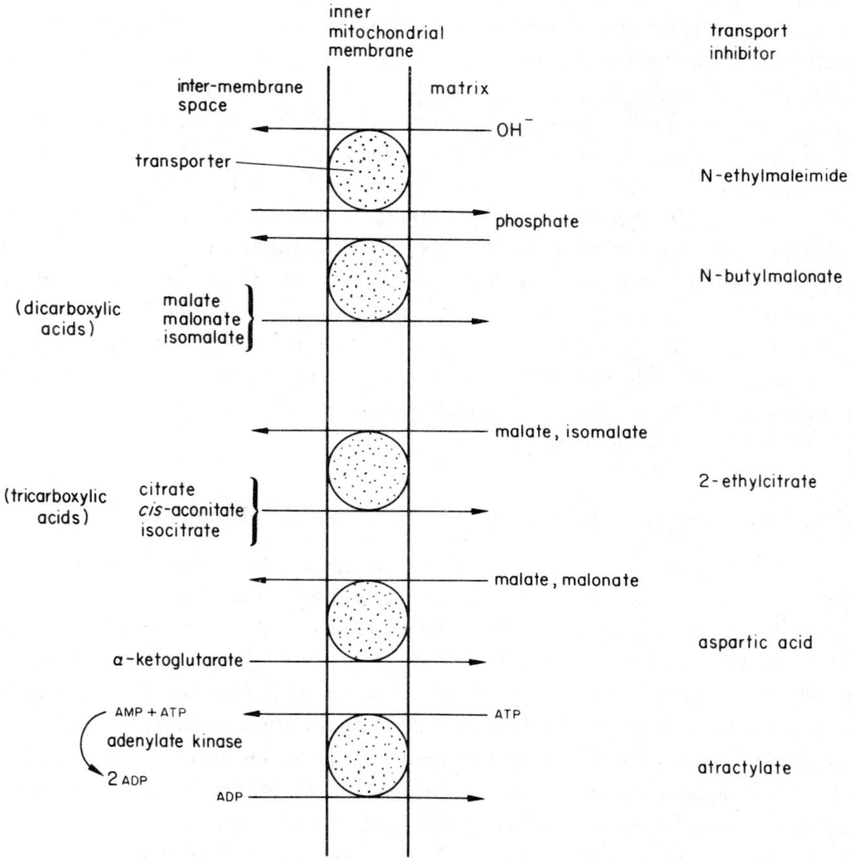

Figure 5.12 Exchange diffusion of anions across the inner membrane of rat-liver mitochondria.

and can be blocked by N-ethyl maleimide or the organomercurial compound mersalyl, which are likely to act by preventing the dissociation of phosphate from the transport site[62]. Once inside the matrix, the phosphate can be used for ATP synthesis or transported out again into the inter-membrane space in exchange for certain dicarboxylic acids (succinate, malate, malonate or isomalate). This exchange-transport process can be specifically blocked by the analogue, N-butyl malonate. Malate or isomalate in the matrix can then be exchanged, in a process sensitive to 2-ethyl citrate, for the tricarboxylic acid anions, citrate, isocitrate or cis-aconitate. An aspartic acid-sensitive carrier-mediated uptake of α-ketoglutarate into the matrix has also been found coupled with the outward movement of malate or malonate. Separate transport systems have been described for certain amino acid anions, such as glutamate and aspartate, but these are not found in mitochondria of all cells. The direction in which substrates are transported across the inner mitochondrial membrane has important consequences in regard to mitochondrial function, since key intermediates between anaerobic and aerobic pathways, such as pyruvate, enter the matrix fairly readily, whereas tricarboxylic acid cycle intermediates are more constrained within the matrix where the enzymes of this pathway are located.

Another exchange-diffusion process of major importance to mitochondrial function is also located in the inner mitochondrial membrane. This involves the obligatory exchange of ATP generated in the matrix compartment for ADP in the inter-membrane space[63]. AMP, which is a product of many reactions within the cell, can be phosphorylated by the enzyme adenylate kinase in the inter-membrane space, which catalyses the reaction: AMP + ATP $\rightleftharpoons$ 2ADP. The nucleotide carrier is specific for ATP/ADP exchange and is specifically inhibited by atractylate, which removes adenine nucleotides from their binding site on the carrier. Other nucleotides, including AMP, have low affinity for the transport site. The net movement of ATP out of and ADP into the matrix is facilitated by an electrical potential gradient across the membrane which, if destroyed, prevents the net uptake of ADP into the matrix. The electrical potential gradient is opposed by a pH gradient across the membrane, but on balance the net gradient favours ADP accumulation in the matrix.

5.3.3 Cation Transport Across the Inner Mitochondrial Membrane

Metabolic intermediates and nucleotide phosphates are anionic in character and, as we have seen, passage through the inner membrane is related to the charge carried by the permeant ions. Similar restrictions also apply to cations that cannot diffuse freely through the membrane, a property that is thought to be intimately concerned with the mechanism of oxidative phosphorylation. Certain inorganic cations, including protons, however, can be transported across the inner mitochondrial membrane against their respective concentration gradients in a manner closely linked to respiration (figure 5.13). In the presence of phosphate, for example, respiring mitochondria can take up calcium, strontium or manganese (but not magnesium) and the corresponding phosphate salts accumulate in the matrix. Protons are ejected from the matrix concomitantly with cation uptake, and under these conditions there is no net ATP synthesis; most agents (with the notable exception of oligomycin) that prevent ATP synthesis, however, also block cation accumulation.

The role of selective cation permeability in exercising control over mitochondrial respiration can be demonstrated by noting the consequences of the destruction of cation gradients across the membrane. There is a loss of respiratory control, ATP synthesis stops and phosphorylation is uncoupled from oxidation–reduction reactions, which, judging from the increased oxygen consumption, usually accelerate. The action of specific agents capable of uncoupling oxidation from phosphorylation can also be explained in terms of their ability to disrupt cation gradients across the inner mitochondrial membrane. The uncoupling agent, 2,4-dinitrophenol, in common with certain other such agents, is a lipophilic weak acid, which is thought to take up protons from the inter-membrane compartment and transport them along a concentration gradient back into the matrix[64] (see section 4.2.3). Cyclic peptide antibiotics such as valinomycin and gramicidin also uncouple oxidative phosphorylation, but in this case potassium ions are conducted into the matrix when this cation is in higher concentration in the inter-membrane region.

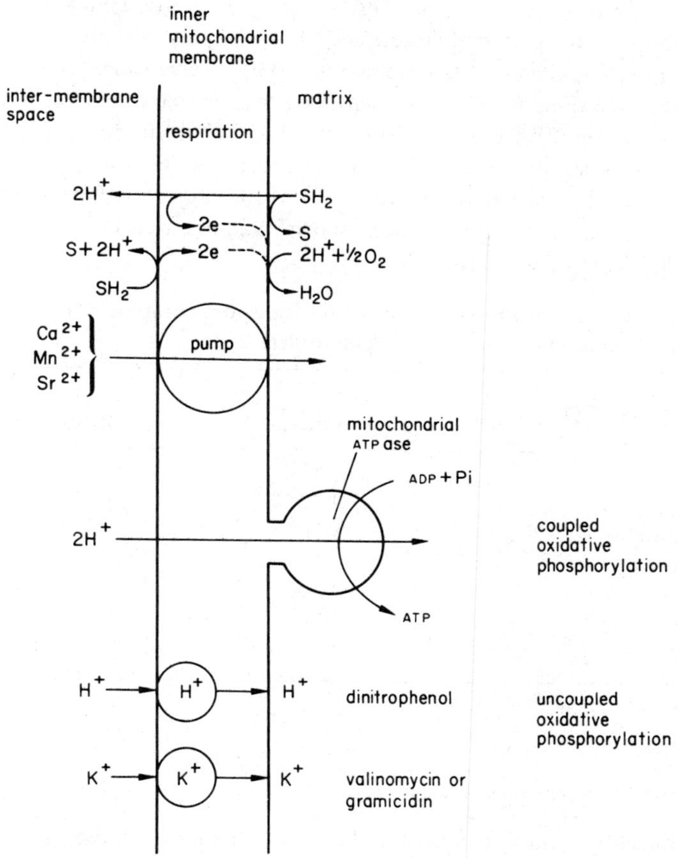

Figure 5.13 Cation transport across the inner mitochondrial membrane. The movement of cations across the inner mitochondrial membrane as postulated by the chemiosmotic hypothesis for oxidative phosphorylation coupling is shown. During respiration, protons are transported out of the matrix and electrons inwards by components of the electron-transport chain. A total of six protons are ejected during the passage of a pair of electrons from reduced substrate ($NADH_2$) to molecular oxygen and three ADP molecules are phosphorylated; thus the return of two protons to the matrix supplies sufficient energy for each coupled phosphorylation reaction. Ca^{2+}, Mn^{2+} or Sr^{2+} can be taken up into the matrix; no net ATP synthesis occurs under these conditions. Phosphorylation is uncoupled from oxidation by indiscriminate transport of protons or potassium ions into the matrix.

5.3.4 The Site of Oxidative Phosphorylation

Oxidative phosphorylation is the process linking the transfer of energy derived from the oxidation of reduced substrates to the terminal phosphate ester bond of ATP. All the reactions catalysing this transfer in animal cells are located exclusively in the inner membrane of mitochondria. A substantial proportion of the constituents of this membrane appear to be involved in some way with oxidative phosphorylation and it has been calculated that approximately a quarter of membrane proteins are concerned directly in these reactions. The relative amounts of various individual components, some of which are shown in table 5.1, indicate that electron carriers and phosphorylation enzymes, together with their associated coupling factors and phospholipids, are integrated into structurally inter-dependent functional units. The mechanism of energy conservation in ATP synthesis is not known. Two theories have been formulated, however, to account for this, and both have their ardent proponents. While not intending to debate the individual merits of these theories (for this purpose the reader is referred to a comprehensive review of the subject by Greville[66]) a discussion of the way the inner mitochondrial membrane is thought to be involved seems to be warranted.

Table 5.1 Relative proportion of electron-transport components in the inner membrane of rat-liver mitochondria

Respiratory carrier	$\frac{\mu\text{moles}}{\text{g protein}}$	$\frac{\mu\text{moles}}{\mu\text{mole cytochrome c}}$
NAD^+	3.80	19
flavoprotein	0.72	3.6
ubiquinone	2.10	10.5
cytochrome b	0.18	0.9
cytochrome $c+c_1$	0.34	1.7
cytochrome a	0.20	1.0
cytochrome a_3	0.22	1.1

data collated by Hoch[65]

5.3.5 The Chemical Hypothesis

The chemical hypothesis is based on the same principles as those pertaining to substrate-level phosphorylation, which take place in solution rather than in membrane-associated reactions. Although the site of oxidative phosphorylation is not disputed, the permeability properties of the membrane are not considered essential to the proposed reaction mechanism. This theory postulates that when an electron carrier in a reduced state becomes oxidised by the subsequent component in the electron-transport chain, the energy made available by the oxidation is used to form a covalent bond in a separate nonreduceable component. This bond is then thought to be transferred through one or more coupled reactions to a phosphorylated precursor for the final reaction with ADP to form ATP as follows

$$\begin{aligned} A_{red} + B_{ox} + I &\rightleftharpoons A_{ox} \sim I + B_{red} \\ A_{ox} \sim I + X &\rightleftharpoons A_{ox} + X \sim I \\ X \sim I + P &\rightleftharpoons X \sim P + I \\ X \sim P + \text{ADP} &\rightleftharpoons X + \text{ATP} \end{aligned}$$

where A and B are adjacent respiratory carriers in reduced (red) or oxidised (ox) form at a given coupling site, and X and I are energy-transfer carriers. The overall change in free energy from reduced nicotinamide nucleotide to oxygen can be subdivided into three stages, based on the redox potentials of each of the individual electron-transport components, where sufficient energy could be derived to phosphorylate ADP. Apart from lack of success in isolating postulated high-energy intermediates, which in itself does not necessarily detract from the merits of the hypothesis, the scheme is apparently unable to give a satisfactory account of the uncoupling effect that results from a disruption of the selective permeability of the inner mitochondrial membrane. Nevertheless the hypothesis can provide for the action of certain uncoupling agents and respiratory inhibitors, if their effect is interpreted as interference with the production of hypothetical high-energy intermediates or their ability to participate in energy-transfer reactions.

5.3.6 The Chemiosmotic Theory

The second hypothesis of coupling ATP synthesis to electron transport was first proposed by Mitchell[67] and is known as the chemiosmotic theory. According to this theory, the organisation of inner-membrane components and the permeability characteristics of the membrane are both essential to the coupling mechanism. The dehydrogenase enzymes and redox carriers are believed to be spatially arranged within the membrane so that a directional movement of protons out of, and electrons into, the matrix takes place, and these gradients are preserved by the relative impermeability of the membrane to the ejected protons. The protons extruded during respiration are thought to return to the matrix across a potential difference created by their outward movement (estimated to be about 250 mV) in such a way that the energy can be utilised for ATP synthesis. In this process, protons localised in the membrane matrix are believed to drive the reaction

$$\begin{matrix} \diagdown \vert \\ \;\;\text{P—OH} \\ \diagup \vert \end{matrix} + H^+ \rightleftharpoons \begin{matrix} \vert \\ -P^{+}- \\ \vert \end{matrix} + H_2O$$

which is required for the subsequent condensation step in pyrophosphate-bond formation. Respiratory control is achieved by a tight coupling between inward proton movement and phosphorylation of ADP. Relief from respiratory control by uncoupling agents is explained by indiscriminate movement of protons into the matrix or, in the case of the cyclic peptide antibiotics, destruction of the potential gradient across the membrane. This theory, in contrast to the chemical hypothesis, actually predicts that destruction of selective permeability of the inner mitochondrial membrane will uncouple oxidation from phosphorylation. The basic tenets of the chemiosmotic hypothesis therefore imply that components of the respiratory chain must be arranged so as to achieve a vectoral movement of electrons and protons across the membrane. There is some experimental verification that this is indeed the case, although the disposition of all the respiratory components must be established before the theory can be properly validated. A possible arrangement of these components within the inner membrane that may satisfy many of the postulates of the chemiosmotic theory are depicted schematically in figure 5.14.

Studies of the molecular topography of the inner mitochondrial membrane have been conducted using a number of different methods, but all indicate that

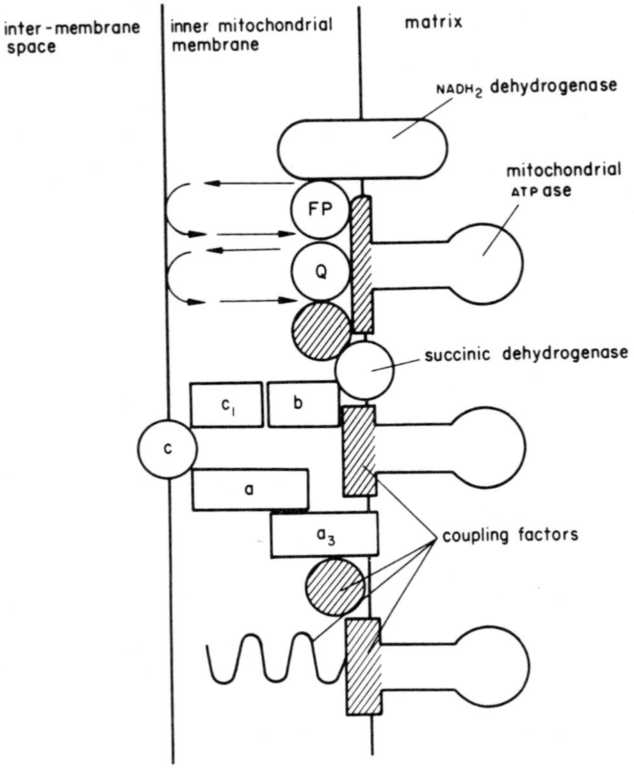

Figure 5.14 Diagram of a possible arrangement of the components of oxidative phosphorylation in the inner mitochondrial membrane. Only the relative disposition of the components is illustrated; the proportions of flavines (FP), ubiquinones (Q), cytochrome b (b), cytochromes c and c_1 ($c + c_1$) and cytochrome oxidase (a, a_3) are presented in table 5.1. Note the presence of coupling factors connecting electron transport with ATPase.

the membrane is asymmetric, with some components located on the outer surface, while others are found exclusively at the matrix surface. Histochemical techniques have already been mentioned as a method used to localise the site at which cytochrome c is oxidised by cytochrome oxidase. Another approach has been to extract selectively or purify certain components from the inner mitochondrial membrane and then to define the conditions required for reconstituting a functional system, whether this be electron transport or coupled oxidative phosphorylation. One of the first experiments of this type was reported by Jacobs and Sanadi[68], who depleted intact mitochondria of cytochrome c by treatment with 0.15M KCl and then found that they could restore oxidative phosphorylation simply by adding cytochrome c to extracted mitochondria suspended in low salt medium. Subsequent experiments by other workers showed that cytochrome c could not be extracted from inverted submitochondrial particles unless a preliminary detergent treatment was used to disrupt the membrane, thus confirming that cytochrome c is located on the outer surface of the inner mitochondrial membrane. Racker *et al.*[69] showed that succinate dehydrogenase and several coupling factors could be extracted from the matrix surface of the inner membrane with silicotungstic acid; they were also able to reconstitute a membrane capable of oxidative phosphorylation when these were added back to the depleted particle. The location of one of these coupling factors, the F_1 mitochondrial ATPase, on the matrix surface has been confirmed by immunological techniques. Thus antibodies prepared against F_1 have been found to interact with inverted inner-membrane preparations but not with intact mitochondria. In contrast, antibody against cytochrome c can inhibit succin-oxidase only when added to intact mitochondria, but antibody against cytochrome oxidase can interact from both sides of the membrane. Finally, chemical labelling techniques similar to those used for other cell membranes (see section 3.4.3) have been applied to mitochondria. The ionised marker diazobenzanesulphonate, which is impermeable to the inner mitochondrial membrane, has been used by Schneider *et al.*[70] to label proteins of intact mitochondria and inverted submitochondrial particles. Cytochrome c and mitochondrial ATPase were completely accessible to the label added to mitochondria and sonicated submitochondrial particles, respectively, whereas cytochrome oxidase, which could be labelled in both membrane preparations, was not completely labelled by the reagent. Thus the specific labelling was found to be six times greater when added to solubilised cytochrome oxidase, suggesting that many reactive groups were masked by their position within the membrane. In additional experiments it was demonstrated that cytochrome a is likely to be exposed on the same surface as cytochrome c, and that cytochrome a_3 probably faces the matrix. It is clear that still more information is needed to complete this picture of inner mitochondrial-membrane topography, but once this problem is solved it should remove some of the uncertainties surrounding the coupling mechanism.

5.4 Summary

The membranes of animal cells are adapted to perform specific functions that are necessary both for the survival of the cell and for its ability to co-operate with other cells to the benefit of the organism as a whole. Cellular responses mediated by polypeptide hormones are initiated by a binding of these molecules to specific receptor sites located on the surface of the plasma membrane and these hormones do not penetrate into the cell. Most of these sites are associated with the membrane-bound enzyme adenylcyclase, which is activated by the binding signal, resulting in an increase in concentration of the intracellular messenger compound, cyclic AMP. The binding site for insulin is not connected with adenylcyclase and its primary action appears to be a modification of the permeability characteristics of the plasma membrane. Insulin can also reduce the intracellular concentration of cyclic AMP in certain cells by activating the phosphodiesterase responsible for the hydrolysis of cyclic AMP or by preventing hormone activation of adenylcyclase.

The plasma membrane possesses genetically determined antigenic sites that are highly complex and specific for strains within a particular species All cells have antigens that are involved primarily in recognition, the histocompatibility antigens, while some antigens such as the blood-group determinants are confined to one cell type. Surface antigenic sites are also concerned with the body's defence against invasion by foreign antigens. Two classes of differentiated lymphocytes. T-cells and B-cells, are the primary instruments used to perform immune responses. The T-cells are responsible for cellular immunity, and proliferate in response to the binding of foreign antigen to membrane sites on the cell surface. Cytotoxicity involves the production of lethal substances and their transfer to target cells. The second form of immunity (humoral immunity) is mediated by B-cells, usually in co-operation with T-cells. The B-cells synthesise and secrete antibodies against foreign antigens and lyse foreign cells by a process involving the binding of antibody and serum complement factors to the plasma membrane of target cells. Antigen-binding to the surface receptors of certain cells can initiate transformation from a resting to a dividing state. In other cases normal cells can be transformed into malignant cells by events initiated at the cell surface.

The endoplasmic reticulum and the Golgi complex are concerned in the synthesis and storage of secretory proteins within the cell. Secretory proteins and certain other proteins for use by the cell are synthesised predominantly on ribosomes attached to the endoplasmic reticulum. Membranes may also be involved on another level to activate amino acids prior to their coding by tRNA, and subsequent incorporation into protein. Proteins destined for secretion move through the cysternae of the endoplasmic reticulum and eventually become concentrated in secretory granules in the region of the Golgi. Enzymes able to modify either the secretory product or vesicle membrane are active during the Golgi phase. Secretory cells differ mainly in the way proteins are stored for secretion: some cells may secrete proteins continuously and possess no stored reserves (albumin from the liver and immunoglobulin from B-cells), some store products in cytoplasmic granules (pancreatic exocrine and endocrine cells) and others in extracellular reserves (thyroid). Membranes undergo certain energy-dependent transformations during secretion, but this is likely to involve a rearrangement of the existing components rather than synthesis of new membrane constituents.

Mitochondria are the site of substrate oxidation and the transfer of energy from reduced products to the terminal phosphate bond of ATP. The inner and outer mitochondrial membranes form two compartments – the inter-membrane space where most anaerobic reactions occur and the matrix where oxidations are performed. The separation of intermediate reaction products and the transfer of common substrates between anaerobic and aerobic metabolic pathways is consistent with the highly selective permeability characteristics of the inner mitochondrial membrane. Transporter molecules are located in this membrane and are responsible for the transport of substrates and ions into the matrix. These movements are linked to the extrusion of products and ions so that the electrical potential across the membrane is not disturbed; this transport is referred to as exchange diffusion. The inner membrane is also the site of oxidative phosphorylation and the electron-transport components, and associated enzymes and coupling factors, are thought to be integrated in a highly organised manner in the inner mitochondrial membrane. The mechanism of coupling between oxidation and phosphorylation is not known but the structure and properties of the inner membrane is believed to play an essential role in this process.

References

1. R. Merrell and L. Glaser. Specific recognition of plasma membranes by embryonic cells. *Proc. natn. Acad. Sci. U.S.A.*, **70** (1973), 2794-8
2. M. Rodbell. Cell surface receptor sites. In: *Current Topics in Biochemistry,* (C.B. Anfinsen, R.F. Goldberger and A.N. Schecter, eds), Academic Press, New York (1972), pp 187-218
3. L. Birnbaumer, S.L. Pohl, H. Michiel, J. Krans and M. Rodbell. The actions of hormones on the adenylcyclase system. In: *Role of Cyclic AMP in Cell Function,* Vol. 3, (P. Greengard and E. Costa, eds), Advances in Biochemical Psychopharmacology. Raven Press, New York (1970), pp 185-208
4. P.W. Kreiner, J.J. Keirns and M.W. Bitensky. A temperature-sensitive change in the energy of activation of hormone-stimulated hepatic adenylcyclase. *Proc. natn. Acad. Sci. U.S.A.*, **70** (1973), 1785-9
5. G. Puchwein, T. Pfeuffer and E.J.M. Helmreich. Uncoupling of catecholamine activation of pigeon erythrocyte membrane adenylate cyclase by filipin. *J. biol. Chem.*, **249** (1974), 3232-40
6. I.H. Pastan. Current directions in research on cyclic AMP. In: *Current Topics in Biochemistry,* (C.B. Anfinsen, R.F. Goldberger and A.N. Schechter eds), Academic Press, New York (1972), pp 65-100
7. G.S. Levey. Solubilization of myocardial adenylcyclase: loss of hormone responsiveness and activation by phospholipids. *Ann. N.Y. Acad. Sci.*, **185** (1971), 448-57
8. A. Rethy, V. Tomasi, A. Trevisani and O. Barnabei. The role of phosphatidylserine in the hormonal control of adenylcyclase of rat liver plasma membranes. *Biochim. Biophys. Acta,* **290** (1972), 58-69
9. G.V. Marinetti, L. Schlatz and K. Reilly. Hormone-Membrane interactions. In: *Insulin Action,* (I.B. Fritz, ed.), Academic Press, London (1972), pp 207-39
10. P. Cuatrecasus. Membrane receptors. *A. Rev. Biochem.*, **43** (1974), 169-214
11. H.T. Narahara. Binding of insulin to tissues in relation to biological action of the hormone. In: *Handbook of Physiology,* Section 7, Vol. I, American Physiological Society (1972), pp 333-45
12. D.L. Mann and J.L. Fahey. Histocompatibility antigens. *A. Rev. Microbiol.*, **25** (1971), 679-710
13. R.A. Reisfeld and B.D. Kahan. Human histocompatibility antigens. In: *Contemporary Topics in Immunochemistry,* Vol. I. (F.P. Inman, ed.), Plenum Press, New York (1972), pp 51-76
14. W.M. Watkins. Blood-group substances. *Science, N.Y.*, **152** (1966), 172-81
15. F.A. Green. Erythrocyte membrane lipids and Rh antigen activity. *J. biol.*

Chem., **247** (1972), 881–7

16. M.C. Raff. T and B lymphocytes and immune responses. *Nature, Lond.*, **242** (1973), 19–23
17. J.F.A.P. Miller. Lymphocyte interactions in antibody responses. *Int. Rev. Cytol.*, **33** (1972), 77–130
18. M.J. Taussig. T-cell factor which can replace T-cells *in vivo*. *Nature Lond.*, **248** (1974), 234–6
19. M. Zembala, W. Ptak and M. Hanczakowska. The role of macrophages in the cytotoxic killing of tumour cells *in vitro*. 1. Primary immunization of lymphocytes *in vitro* for target cell killing and the mechanism of lymphocyte-macrophage co-operation. *Immunology*, **25** (1973), 631–44
20. A. Temple, G. Loewi, P. Davies and A. Honard. Cytotoxicity of immune guinea-pig cells. II. The mechanism of macrophage cytotoxicity. *Immunology*, **24** (1973), 655–69
21. H.J. Müller-Eberhard, *The Molecular Basis of the Biological Activities of Complement*, The Harvey Lectures, Series 66, Academic Press, New York (1972), pp 75–104
22. P.J. Lachmann, D.E. Bowyer, P. Nicol, R.M.C. Dawson and E.A. Munn. Studies on the terminal stages of complement lysis. *Immunology*, **24** (1973), 135–45
23. M. Greaves and G. Janossy. Elicitation of selective T and B lymphocyte responses by cell surface binding ligands. *Transplant Rev.*, **11** (1972), 87–130
24. N. Sharon and H. Lis. Lectins: cell-agglutinating and sugar specific proteins. *Science, N.Y.*, **177** (1972), 949–59
25. M. Green. Oncogenic viruses. *A. Rev. Biochem.*, **39** (1970), 701–56
26. M.M. Burger. Cell surfaces in neoplastic transformation. In: *Current Topics in Cellular Regulation*, (B.L. Horecker and E.R. Stadtman eds), Academic Press, London (1971), pp 135–93
27. J.R. Tata. Co-ordinated synthesis of membrane phospholipids with the formation of ribosomes and accelerated protein synthesis during growth and development. In: *Current Trends in the Biochemistry of Lipids*, (J. Ganguly and R.M.S. Smellie, eds), Academic Press, London (1972), pp 333–46
28. T.M. Andrews and J.R. Tata. Protein synthesis by membrane-bound and free ribosomes of secretory and non-secretory tissues. *Biochem. J.*, **121** (1971), 683–94
29. Y. Ikehara and H.C. Pitot. Localization of polysome-bound albumin and serine dehydratase in liver cell fractions. *J. Cell Biol.*, **59** (1973), 28–44
30. M. Rosbash. Formation of membrane-bound polyribosomes. *J. molec. Biol.*, **65** (1972), 413–22
31. F.S. Rolleston. Membrane-bound and free ribosomes. *Sub-cell. Biochem.*, **3** (1974), 91–117
32. T.K. Shires, T. Ekren, L.M. Narurkar and H.C. Pitot. Protein synthesis on

rat liver polysome–membrane complexes formed *in vitro* and disposition of the discharged chains. *Nature new Biol.*, **242** (1973), 198–201

33. M.R. Adelman *et al.* 1. An improved fractionation procedure for the preparation of rat liver membrane-bound ribosomes. 2. Ribosome–membrane interaction. *J. Cell Biol.*, **56** (1973), 191–229

34. G.H. Sunshine, D.J. Williams and B.R. Rabin. Role of steroid hormones in the interaction of ribosomes with the endoplasmic membranes of rat liver. *Nature new Biol.*, **230** (1971), 133–6

35. C.A. Blyth, R.B. Freedman and B.R. Rabin. Sex specific binding of steroid hormones to microsomal membranes of rat liver. *Nature new Biol.*, **230** (1971), 137–9

36. R.B. Loftfield. The mechanism of aminoacylation of transfer RNA. *Prog. nucl. Acid Res. molec. Biol.*, **12** (1972), 87–128

37. A.K. Bandyopadhyay and M.P. Deutscher. Lipids associated with the aminoacyl-transfer RNA synthetase complex. *J. molec. Biol.*, **74** (1973), 257–61

38. A. Hampel and M.D. Enger. Subcellular distribution of aminoacyl-transfer RNA synthetases in chinese hamster ovary cell culture. *J. molec. Biol.*, **79** (1973), 285–93

39. R.W. Hendler. Lipoamino acids as precursors in the synthesis of authentic proteins by homogenates of hen oviduct. *Proc. natn. Acad. Sci. U.S.A.*, **54** (1965), 1233–40

40. J.D. Jamieson. Transport and discharge of exportable proteins from pancreatic exocrine cells: *in vitro* studies. In: *Current Topics in Membranes and Transport,* Vol. 3, (F. Bronner and A. Kleinzeller, eds), Academic Press, New York (1972), pp 273–338

41. L.G. Caro and G.E. Palade. Protein synthesis, storage and discharge in the pancreatic exocrine cell. An autoradiographic study. *J. Cell Biol.*, **20** (1964), 473–95

42. J.D. Jamieson and G.E. Palade. Intracellular transport of secretory proteins in the pancreatic exocrine cell. I. Role of peripheral elements of the Golgi complex. *J. Cell Biol.*, **34** (1967), 577–96

43. J.D. Jamieson and G.E. Palade. Intracellular transport of secretory proteins in the pancreatic exocrine cell. II. Transport to condensing vacuoles and zymogen granules. *J. Cell Biol.*, **34** (1967), 597–615

44. C.W. Heald and R.G. Saacke. Cytological comparison of milk protein synthesis of rat mammary tissue *in vivo* and *in vitro*. *J. Dairy Sci.*, **55** (1972), 621–8

45. S.L. Howell, M. Kostianovsky and P.E. Lacy. Beta granule formation in isolated islets of Langerhans. A study by electron microscopic radioautography. *J. Cell Biol.*, **42** (1969), 695–705

46. D.F. Steiner, W. Kemmler, J.L. Clark, P.E. Oyer and A.H. Rubinstein. The biosynthesis of insulin. In: *Handbook of Physiology*, Section 7, Vol. 1, American Physiological Society (1972), pp 175–98

47. T. Blundell, G. Dodson, D. Hodgkin and D. Mercoa. Insulin: the structure in the crystal and its reflection in chemistry and biology. *Adv. Protein Chem.*, **26** (1972), 279–402

48. A.E. Kitabachi, W.C. Duckworth, F.B. Stentz and S. Yu. Properties of proinsulin and related polypeptides. *Critical Rev. Biochem.*, **1** (1972), 59–94

49. P.T. Grant and T.L. Coombs. Proinsulin, a biosynthetic precursor of insulin. In: *Essays in Biochemistry,* Vol. 6, (P.N. Campbell and F. Dickens, eds), Academic Press, London (1970), pp 69–92

50. J.B. Stanbury. Some recent developments in the physiology of the thyroid gland. *Rev. Physiol.,* **65** (1972), 94–125

51. T. Onaya and D.H. Solomon. Effects of chlorpromazine and propanolol on *in vitro* thyroid activation by thyrotropin, long-acting thyroid stimulator and dibutyryl cyclic-AMP. *Endocrinology,* **85** (1969), 1010– 7

52. J.D. Jamieson and G.E. Palade. Intracellular transport of secretory proteins in the pancreatic exocrine cell. III. Dissociation of intracellular transport from protein synthesis. *J. Cell Biol.,* **39** (1968), 580–8

53. M. Dauwalder, W.G. Waley and J.E. Kephart. Functional aspects of the Golgi apparatus. *Sub-cell Biochem.,* **1** (1972), 225–75

54. J.D. Jamieson and G.E. Palade. Intracellular transport of secretory proteins in the pancreatic exocrine cell. IV. Metabolic requirements. *J. Cell Biol.,* **39** (1968), 589–603

55. J.J.M. Bergeron, J.H. Ehrenreich, P. Siekevitz and G.E. Palade. Golgi fractions prepared from rat liver homogenates. II. Biochemical characterisation. *J. Cell Biol.*, **59** (1973), 73–88

56. H. Rasmussen. Cell communication, calcium ion and cyclic adenosine monophosphate. *Science, N.Y.,* **170** (1970), 404–12

57. K.A. Gumaa, P. McLean and A.L. Greenbaum. Compartmentation in relation to metabolic control in liver. In: *Essays in Biochemistry,* Vol. 7, (P.N. Campbell and F. Dickens, eds), Academic Press, London (1971), pp 39–86

58. A.M. Seligman, M.J. Karnovsky, H.L. Wasserkrug and J.S. Hanker. Nondroplet ultrastructural demonstration of cytochrome oxidase activity with a polymerising osmiophilic reagent, diaminobenzidine (DAB). *J. Cell Biol.,* **38** (1968), 1–14

59. E. Racker, C. Burstein, A. Loyter and R.O. Christiansen. The sidedness of the inner mitochondrial membrane. In: *Electron Transport and Energy Conservation,* (J.M. Tager, S. Papa, E. Quagliariello and E.C. Slater, eds), Adriatica Editrice, Bari (1970), pp 235–48

60. E.J. Harris and J.R. Manger. Intersubstrate competitions and evidence for compartmentation in mitochondria. *Biochem. J.,* **113** (1969), 617–28

61. J.B. Chappell, J.D. McGivan and M. Crompton. The anion transporting systems of mitochondria and their biochemical significance. In: *The Molecular Basis of Biological Transport,* (J.F. Woessner and F. Huijing, eds),

Academic Press, London (1972), pp 55–81

62. N.E. Lofrumento, S. Papa, F. Zanotti and E. Quagliariello. On the inhibition of the transport of inorganic phosphate by N-ethyl-maleimide. *FEBS Letters,* **36** (1973), 273–6

63. M. Klingenberg. Metabolite transport in mitochondria: an example for intracellular membrane function. In: *Essays in Biochemistry,* Vol. 6, (P.N. Campbell and F. Dickens, eds), Academic Press, London (1970), pp 119–59

64. P. Mitchell and J. Moyle. Estimation of membrane potential and pH difference across the cristae membrane of rat liver mitochondria. *Eur. J. Biochem.,* 7 (1969), 471–84

65. F. Hoch. *Energy Transformations in Mammals: Regulatory Mechanisms,* Saunders, Philadelphia (1971)

66. G.D. Greville. A scrutiny of Mitchell's chemiosmotic hypothesis of respiratory chain and photosynthetic phosphorylation. In *Current Topics in Bioenergetics,* Vol. 3, (D.R. Sanadi, ed.). Academic Press, New York (1969), pp 1–78

67. P. Mitchell. Chemiosmotic coupling in oxidative and photosynthetic phosphorylation. *Biol. Rev.,* **41** (1966), 445-502

68. E.E. Jacobs and D.R. Sanadi. The reversible removal of cytochrome c from mitochondria. *J. biol. Chem.,* **235** (1960), 531–4

69. E. Racker, LL. Horstman, D. Kling and J.M. Fessenden-Raden. Partial Resolution of the enzyme catalysing oxidative phosphorylation. XXI. Resolution of submitochondrial particles from bovine heart mitochondria with silicotungstate. *J. biol. Chem.,* **244** (1969), 6668–74

70. D.L. Schneider, Y. Kagawa and E. Racker. Chemical modification of the inner mitochondrial membrane. *J. biol. Chem.,* **247** (1972), 4074–9

Index

BCL3